Studies in Applied Philosophy, Epistemology and Rational Ethics

Volume 54

Studies in Applied Philosophy, Epistemology and Rational Ethics (SAPERE) publishes new developments and advances in all the fields of philosophy, epistemology, and ethics, bringing them together with a cluster of scientific disciplines and technological outcomes: ranging from computer science to life sciences, from economics, law, and education to engineering, logic, and mathematics, from medicine to physics, human sciences, and politics. The series aims at covering all the challenging philosophical and ethical themes of contemporary society, making them appropriately applicable to contemporary theoretical and practical problems, impasses, controversies, and conflicts. Our scientific and technological era has offered "new" topics to all areas of philosophy and ethics – for instance concerning scientific rationality, creativity, human and artificial intelligence, social and folk epistemology, ordinary reasoning, cognitive niches and cultural evolution, ecological crisis, ecologically situated rationality, consciousness, freedom and responsibility, human identity and uniqueness, cooperation, altruism, intersubjectivity and empathy, spirituality, violence. The impact of such topics has been mainly undermined by contemporary cultural settings, whereas they should increase the demand of interdisciplinary applied knowledge and fresh and original understanding. In turn, traditional philosophical and ethical themes have been profoundly affected and transformed as well: they should be further examined as embedded and applied within their scientific and technological environments so to update their received and often old-fashioned disciplinary treatment and appeal. Applying philosophy individuates therefore a new research commitment for the 21st century, focused on the main problems of recent methodological, logical, epistemological, and cognitive aspects of modeling activities employed both in intellectual and scientific discovery, and in technological innovation, including the computational tools intertwined with such practices, to understand them in a wide and integrated perspective.

Studies in Applied Philosophy, Epistemology and Rational Ethics means to demonstrate the contemporary practical relevance of this novel philosophical approach and thus to provide a home for monographs, lecture notes, selected contributions from specialized conferences and workshops as well as selected Ph.D. theses. The series welcomes contributions from philosophers as well as from scientists, engineers, and intellectuals interested in showing how applying philosophy can increase knowledge about our current world. Initial proposals can be sent to the Editor-in-Chief, Prof. Lorenzo Magnani, lmagnani@unipv.it:

- A short synopsis of the work or the introduction chapter
- The proposed Table of Contents
- The CV of the lead author(s).

For more information, please contact the Editor-in-Chief at lmagnani@unipv.it.

Indexed by SCOPUS, ISI and Springerlink. The books of the series are submitted for indexing to Web of Science.

More information about this series at http://www.springer.com/series/10087

Adriano Fabris
Editor

Trust

A Philosophical Approach

Springer

Editor
Adriano Fabris
Pisa University
Pisa, Italy

ISSN 2192-6255 ISSN 2192-6263 (electronic)
Studies in Applied Philosophy, Epistemology and Rational Ethics
ISBN 978-3-030-44017-6 ISBN 978-3-030-44018-3 (eBook)
https://doi.org/10.1007/978-3-030-44018-3

This Springer imprint is published by the registered company Springer Nature Switzerland AG
The registered company address is: Gewerbestrasse 11, 6330 Cham, Switzerland

Organization

Steering Committee

General Chair

Adriano Fabris, Pisa University, Italy

Co-Chair

Roberto Gronda, Pisa University, Italy

Members

Leonardo Amoroso, Pisa University, Italy
Pierluigi Barrotta, Pisa University, Italy
Raimondo Cubeddu, Pisa University, Italy
Cristina D'Ancona, Pisa University, Italy
Fabio Dei, Pisa University, Italy
Enrico Moriconi, Pisa University, Italy
Veronica Neri, Pisa University, Italy

Program Committee

Leonardo Amoroso, Pisa University, Italy
Pierluigi Barrotta, Pisa University, Italy
Elisa Coda, Pisa University, Italy
Raimondo Cubeddu, Pisa University, Italy
Cristina D'Ancona, Pisa University, Italy
Fabio Dei, Pisa University, Italy
Caterina Di Pasquale, Pisa University, Italy
Antonio Masala, Pisa University, Italy
Marco Menon, Pisa University, Italy
Enrico Moriconi, Pisa University, Italy

Preface

This book is the crowning achievement of a research into "trust", coordinated by a research team of the University of Pisa. As suggested by the subtitle, "trust" has been mainly addressed from a philosophical point of view. However, in this book, the philosophical approach, developed in historical and systematic forms, was related in an interdisciplinary way with exact sciences, natural sciences and social sciences (especially communication sciences and political sciences).

Trust—who should be trusted and what deserves trust, how a community is built on mutual trust, and how trust can be strengthened or undermined—has become ever more important nowadays, in a time in which the mechanisms of such human attitude have dramatically changed. Generally speaking, by "trust" we mean the ability to trust or rely on someone or something, on a human or a process, on a situation or—in a religious context—on God. *In God We Trust,* as we know, is the national motto of the United States of America.

There is a reason for such confidence. Often, we cannot perfectly control some contexts of our lives or some mechanisms we have to do with. So, we rely on someone or something that seems to be able to control that or seems to act in a predictable way. It is someone or something that, with good reason and because it can be verified, seems trustworthy. So, until proven otherwise, we can trust it.

However, the very structure of trust has changed nowadays. Processes that even in a recent past turned out to be certainly reliable are no longer regarded as such now. We question many forms of knowledge and many of its results, including those of scientific knowledge. For instance, we no longer confidently rely on the experts who used to vouch for such knowledge. In other words, it seems that in today's world distrust has generally grown, even distrust for issues that were not questioned before.

At the same time, along with a decline of trust, we see a bewildering swell in fields and objects that used to be regarded as reliable. I am speaking, for instance, of all that the technological devices give access to. I am speaking of the operation of such devices. All this has consequences on the operation of the institutions, on social cohesion, on the forms of politics we have to do with now as well.

The proliferation of the so-called "populisms" can actually be associated with a change in the ways trust is given by voters to rulers.

That's why nowadays we should reflect again on the matter of trust and its transformations. This is the subject that the research project "Ethics, Science, Democracy" funded by the University of Pisa in 2018 through the PRA programme (Progetti di Ricerca di Ateneo: University Research Projects) has been focussed on. It has investigated some specific issues that are related to the underlying matter, first and foremost in some of the papers that have been published in "Teoria" ("Teoria" 2019/1 and 2019/2: *The Prismatic Shape of Trust; 1. A Theoretical Approach; 2. Authors and Problems*). Then, and above all, the PRA research project came into its own and found its highest point in an international conference called *Trust. A Philosophical Approach,* held at the University of Pisa on May 11 and 12, 2020.

It is the same title as that of the book. Like the conference, this book too consists of four topical parts. From the perspective of the history of philosophy, the first part investigates some of the moments in the antique, late antique, and medieval worlds in which the issue of trust was developed, based on a specific concept of knowledge as well as on an experience of faith, sometimes in conflict with knowledge itself. The second part delves deeper, from the perspective of the philosophy of science, into some of the ways trust is built and kept, through scientific arguments and evidence, while also looking at the developments and consequences, the transformations and changes in such approach. The third part takes an interdisciplinary approach to deal with the changes that the bond of trust has gone and is going through, in the technological and communicative sphere, and provides a discussion of the ontological status of such concept. The fourth and last part investigates the relationship between trust and political institutions, and how such bond is changing in today's world.

Leonardo Ceragioli and Marta Vero too, as research fellows, joined for some time in the PRA research project and in the thematic insights that have been accomplished herein.

In addition to the speakers whose papers have been published herein, Mauro Mariani too took part in the conference with the paper *Is Aristotelian Deduction Still Reliable?,* which unfortunately could not be ready early enough to be included in this book.

Along with the members of the Programme Committee, I would like to thank the invited external reviewers, who reviewed these papers thoroughly, knowledgeably, and unwaveringly. Then, my special thanks go to Roberto Gronda, who assisted me in the organization of the conference and throughout the publication of this book And lastly, I would like to express my gratitude to the University of Pisa for its generous support and show all my appreciation for the partnership with Springer, and especially for Leontina Di Cecco's invaluable help.

Pisa, Italy Adriano Fabris

Contents

History of Philosophy

Aristotle on Scientific Principles (*Posterior Analytics* 1. 1–2)

Paolo Crivelli

Abstract In the first book of the *Posterior Analytics*, Aristotle distinguishes two types of preliminary knowledge that can be involved in scientific understanding: knowing *that* something is and knowing *what* something is, which he treats as equivalent to knowing what something signifies. What do these two types of preliminary knowledge amount to? How are they reciprocally related? On what grounds may they be regarded as trustworthy? It turns out that knowing that something *is* probably covers three types of knowledge: knowing that something *is the case*, knowing that something *exists*, and knowing that something *is so-and-so*. Given this wide spectrum covered by knowing *that* something is, the problem arises of understanding what logical space is left for knowing *what* something is. Unless knowing *what* something is a particular case of knowing *that* something is, the most plausible guess is that knowledge of *what* something is, is knowledge whose linguistic formulation involves the 'definitional' reading of sentences, namely the reading of a sentence whereby its predicate-expression is taken to offer a complete description of the nature of the kind signified by its subject-expression. The trustworthiness of our grasp of definitions is due to a special faculty postulated by Aristotle, namely the intellect, which has the power to 'look into' the nature of the kinds it is concerned with.

Keywords Aristotle · Conviction · Definition · Existence · Knowledge · Posit · Predication · Principle · Supposition · Trustworthiness · Truth

P. Crivelli (✉)
Université de Genève, Genève, Switzerland
e-mail: Paolo.Crivelli@unige.ch

A. Fabris (ed.), *Trust*, Studies in Applied Philosophy, Epistemology and Rational Ethics 54, https://doi.org/10.1007/978-3-030-44018-3_1

In the first book of the *Posterior Analytics*, Aristotle distinguishes two types of preliminary knowledge that can be involved in scientific understanding: knowing *that* something is and knowing *what* something is, which he treats as equivalent to knowing what something signifies. In this study I investigate what these two types of preliminary knowledge amount to, how are they reciprocally related, and on what grounds they may be regarded as trustworthy.

It turns out that knowing that something *is* probably covers three types of knowledge: knowing that something *is the case*, knowing that something *exists*, and knowing that something *is so-and-so*. Given this wide spectrum covered by knowing *that* something is, the problem arises of understanding what logical space is left for knowing *what* something is. Unless knowing *what* something is a particular case of knowing *that* something is, the most plausible guess is that knowledge of *what* something is, is knowledge whose linguistic formulation involves the 'definitional' reading of sentences, namely the reading of a sentence whereby its predicate-expression is taken to offer a complete description of the nature of the kind signified by its subject-expression. The trustworthiness of our grasp of definitions is due to a special faculty postulated by Aristotle, namely the intellect, which has the power to 'look into' the nature of the kinds it is concerned with.

The structure of the first two chapters of the Posterior Analytics. In the first two chapters of the *Posterior Analytics* Aristotle lays the foundations of his philosophy of science. The basis of Aristotle's discussion in these two chapters is formulated in the opening sentence of the treatise:

T1 Πᾶσα διδασκαλία καὶ πᾶσα μάθησις διανοητικὴ ἐκ προϋ- 71ᵃ1
παρχούσης γίνεται γνώσεως. 71ᵃ2

All teaching and all discursive learning[1] come about from already existing knowledge. (Arist. *APo.* 1.1, 71ᵃ1–2)

The first two chapters of the *Posterior Analytics* revolve around the thesis formulated in T1.

The first chapter can be divided into four sections connected with the thesis that all teaching and all discursive learning[1] come about from already existing knowledge: (1) a formulation and a brief justification of the thesis itself (71ᵃ1–11); (2) a distinction of two types of already existing knowledge (71ᵃ11–17); (3) the claim that only part of the already existing knowledge need precede in time the teaching or discursive learning (71ᵃ17–24); (4) the claim that there is a sense in which it can be said that one already knows what one is being taught or discursively learning (71ᵃ24–ᵇ8). The second chapter introduces the concept of understanding (ἐπίστασθαι) and links it with the thesis that all teaching and all discursive learning come about from already existing knowledge. The chapter can be divided into three sections: (1) an account of the nature of understanding which entails that the type of understanding

[1]Discursive learning could be the acquisition of knowledge based on reasoning and argument (opposed to the acquisition of knowledge based on perception) or the acquisition of knowledge of facts (opposed to the acquisition of knowledge of things).

on which the *Posterior Analytics* concentrate is demonstration, i.e. valid deductive argument whose ultimate premisses enjoy certain characteristics (in particular, they are true, primitive, non-demonstrable, explanatory of the conclusion, and prior to it) (71^b9–72^a5); (2) a classification of the types of ultimate demonstrative premisses, or principles (72^a5–24); (3) a discussion of how a demonstrator must be convinced of the demonstrative principles and what is demonstrated from them (72^a25–$^b4$). The examination of the nature of understanding and demonstration is linked to that of the knowledge that precedes all teaching and all discursive learning because demonstrative principles are among the things whose knowledge precedes that of what gets demonstrated and the process of demonstrating is central to teaching and discursive learning.

Two types of preliminary knowledge. Here is the second section of the first chapter of the *Posterior Analytics*, where a distinction of two types of already existing knowledge is drawn:

T2 διχῶς δ᾽ ἀναγκαῖον προγινώσκειν· τὰ μὲν γάρ, ὅτι ἔστι, προϋπολαμβάνειν ἀναγκαῖον, τὰ δέ, τί τὸ λεγόμενόν ἐστι, ξυνιέναι δεῖ, τὰ δ᾽ ἄμφω, οἷον ὅτι μὲν ἅπαν ἢ φῆσαι ἢ ἀποφῆσαι ἀληθές, ὅτι ἔστι, τὸ δὲ τρίγωνον, ὅτι τοδὶ σημαίνει, τὴν δὲ μονάδα ἄμφω, καὶ τί σημαίνει καὶ ὅτι ἔστιν· οὐ γὰρ ὁμοίως τούτων ἕκαστον δῆλον ἡμῖν.

It is necessary already to know things in two ways: of some things it is necessary to believe already that they are, of some one must grasp what the thing said is, and of others both. For instance, of the fact that it is true to affirm or deny everything ‹one must already believe› that it is; of the triangle, that it signifies this; and of the unit both (both what it signifies and that it is). For each of these is not clear to us in the same way. (Arist. *APo.* 1.1, 71^a11–17)

In passage T2 Aristotle seems to consider one's earlier knowledge of what the unit signifies as an example of earlier knowledge of what the thing said is. A similar connection between signification and what a thing is occurs also in *Posterior Analytics* 2.10, where Aristotle says:

T3 Ὁρισμὸς δ᾽ ἐπειδὴ λέγεται εἶναι λόγος τοῦ τί ἐστι, φανερὸν ὅτι ὁ μέν τις ἔσται λόγος τοῦ τί σημαίνει τὸ ὄνομα ἢ λόγος ἕτερος ὀνοματώδης, οἷον τί σημαίνει [τί ἐστι] τρίγωνον.

Since a definition is said to be an account of the what it is, it is clear that one ‹type of definition› will be an account of what a name or another name-like phrase signifies, for instance what triangle signifies. (Arist. *APo.* 2.10, 93^b29–32)

The joint mention of the ‘what it is’ and signification links passage T3 with passage T2. However, there might be a difference in the use of ‘to signify’ in the two passages: while in T3 Aristotle takes care to specify that what signifies is a linguistic expression (i.e. ‘a name or another name-like phrase’, $93^{b}30$–1), in T2 he might be taken to assume that worldly items have signification (for he says that of the unit one must already believe ‘both what it signifies and that it is’, $71^{a}15$–16, and the unit of which one must already believe that it is is surely not the name ‘unit’ but the unit itself, a worldly item). In fact, while in his treatises that concentrate on logical or semantic questions Aristotle is usually careful about the difference between things in the world and linguistic expressions and applies ‘to signify’ only to linguistic expressions, on some occasions he allows himself a looser usage. For instance, in the *Categories* he says:

T4 Πᾶσα δὲ οὐσία δοκεῖ τόδε τι σημαίνειν. $3^{b}10$

Every substance seems to signify a this-something. (Arist. *Cat.* 5, $3^{b}10$)

(note that not every substance is a linguistic expression). One might be tempted to think that also in passage T2 Aristotle is attributing signification to a worldly item. However, it is perhaps more plausible to assume that in T2 Aristotle allows himself to shift the reference of the expression ‘the unit’: when it is the subject of the predicate ‘is’, the expression ‘the unit’ refers to a worldly item; when it is instead the subject of the predicate ‘signifies’, the expression ‘the unit’ refers to a linguistic expression. Also note that in passage T2 Aristotle speaks of grasping ‘what the thing said is’ ($71^{a}13$): the phrase ‘the thing said’ could be a variant of ‘the words uttered’. The connection between passages T2 and T3 makes it plausible to assume that T2’s already existing knowledge of ‘what the thing said is’ or of what it ‘signifies’ is knowledge whose linguistic expression would be a definition of some sort (recall that in *Posterior Analytics* 2.10 Aristotle distinguishes several types of definition).

Knowing that certain things are*: what use of ‘to be’?* One of the main exegetical problems posed by passage T2 concerns the use of the verb ‘to be’ in the formulation of the claim that one type of already existing knowledge is knowledge that certain things *are*. How is ‘to be’ being used?

In one of his examples, Aristotle indicates that there must be already existing knowledge that the fact that it is true to affirm or deny everything *is* (note that in the Greek original nothing corresponds to the English words ‘the fact’, but it is difficult to find a readable English rendering that does not involve these words). This Aristotelian example unequivocally shows that the use of the verb ‘to be’ in the formulation of the claim that one type of already existing knowledge is knowledge that certain things *are* must at least include the veridical use of the verb, according to which ‘to be’ means ‘to be the case’ or ‘to be true’.

In another of his examples, Aristotle indicates that another case of already existing knowledge that certain things *are* is knowledge that the unit *is*. Whatever the strength of this occurrence of 'to be', it is surely not veridical because the unit is not an item of a propositional nature and the sentences 'The unit is the case' and 'The unit is true' sound ill-formed (unless 'is true' is taken to mean 'is genuine', which however does not fit the other example). The use of the verb 'to be' in the formulation of the claim that there must be earlier knowledge that the unit *is* may be reasonably taken to be either predicative elliptical ('There must be earlier knowledge that the unit is so-and-so') or existential ('There must be earlier knowledge that the unit exists'). A passage in *Posterior Analytics* 1.10, where Aristotle returns to the issue addressed by T2, gives the edge to the existential use. For, in *Posterior Analytics* 1.10 Aristotle says that sciences overlook some of their own principles because these are too obvious to be worth mentioning:

T5 … οὐ γὰρ ὁμοίως δῆλον ὅτι ἀριθμός ἐστι καὶ ὅτι ψυχρὸν $76^{b}18$
καὶ θερμόν. $76^{b}19$

… for it is not equally clear that number is and that the hot and the cold are. (Arist. *APo.* 1.10, $76^{b}18$–19)

Aristotle probably means that while it is obvious that the hot and the cold exist, it is not at all clear that number exists. This suggests that the use of the verb 'to be' in the formulation of the claim about the unit (which is the first of the numbers) is existential. Thus, the use of the verb 'to be' in the formulation of the claim that one type of already existing knowledge is knowledge that certain things *are* probably includes not only the veridical but also the existential use of the verb.

One may wonder about the relationship between the veridical use of 'to be' ('to be' meaning 'to be the case' or 'to be true') and its existential use ('to be' meaning 'to exist'). Do they coincide or are they different? The answer to this question depends on what one takes to be the proper bearers of the attribute expressed by 'to be' on its veridical use according to Aristotle. What is it that, according to Aristotle, can be properly said 'to be the case' or 'to be true'? Two main candidates have been deemed by commentators to compete for such a role: predicative compounds and states of affairs. Predicative compounds are things like walking-Socrates, seated-Theaetetus, and flying-Theaetetus. Walking-Socrates exists just when Socrates is walking, and similarly with seated-Theaetetus and flying Theaetetus. If the proper bearers of the attribute expressed by 'to be' on its veridical use are predicative compounds, then the veridical use of 'to be' comes to coincide with its existential use. States of affairs are entities that can (at least in principle) continue in their existence both when they are the case and when they are not the case (they somewhat resemble propositions as they were conceived of by Carnap). The state of affairs that Socrates is walking exists and is the case when Socrates is walking, and it still exists but is no longer the case when Socrates is not walking. If the proper bearers of the attribute expressed by 'to be' on its veridical use are states of affairs, then the veridical use of 'to be' is distinct from its existential use. It is difficult to reach a decision about what the proper bearers of the attribute expressed by 'to be' on its veridical use are according to Aristotle, and more specifically according to the Aristotle of *Posterior Analytics* 1. I limit myself to observe that the example we find in T2 of a subject for a form of 'to be' on its veridical use, namely 'that it is true to affirm or deny everything'

(71^a13–14), cannot be easily reconciled with predicative complexes (consider the declarative 'that', 'ὅτι', at 71^a13).

In whatever way one settles the question of the relationship between the veridical and the existential use of 'to be', it may be plausibly inferred that both uses are involved in the formulation of the claim that one type of already existing knowledge is knowledge that certain things *are*. However, there are reasons for thinking that this is not all. For, in *Posterior Analytics* 1.10 Aristotle comes back to the issue of the already existing knowledge that certain things *are*. He writes:

T6 Ἔστι δ' ἴδια μὲν καὶ ἃ λαμβάνεται εἶναι, περὶ ἃ ἡ 76^b3
ἐπιστήμη θεωρεῖ τὰ ὑπάρχοντα καθ' αὑτά, οἷον μονάδας ἡ
ἀριθμητική, ἡ δὲ γεωμετρία σημεῖα καὶ γραμμάς. ταῦτα $^b5$
γὰρ λαμβάνουσι τὸ εἶναι καὶ τοδὶ εἶναι. 76^b6

Proper [*sc.* principles] also are the things which are assumed to be, about which the science considers what belongs to them in themselves – as e.g. arithmetic is about units, and geometry is about points and lines. For of these they assume the being and the being this. (Arist. *APo.* 1.10, 76^b3–6)

Passage T6 is ambivalent. For, when, in its last sentence, Aristotle says that 'of these they assume the being and the being this' (76^b5–6), he could mean either of two things: he could mean that the assumption that certain basic items *are* includes both the assumption that they *exists* and the assumption that they *are so-and-so*; or he could mean that with regard to certain items the sciences assume both that they *are* and that they are defined as *being so-and-so*. Aristotle then goes on to contrast the case which he has just been speaking about with that of the per se attributes, with regard to which the science assumes 'what each signifies' (76^b7) and proves that they *are*:

T7 τὰ δὲ τούτων πάθη
καθ' αὑτά, τί μὲν σημαίνει ἕκαστον, λαμβάνουσιν, οἷον ἡ
μὲν ἀριθμητικὴ τί περιττὸν ἢ ἄρτιον ἢ τετράγωνον ἢ κύβος,
ἡ δὲ γεωμετρία τί τὸ ἄλογον ἢ τὸ κεκλάσθαι ἢ νεύειν, ὅτι
δ' ἔστι, δεικνύουσι διά τε τῶν κοινῶν καὶ ἐκ τῶν ἀποδεδει-
γμένων.

As for the *per se* affiections of these, they [*sc.* the various sciences] assume what each signifies, for instance arithmetic ‹assumes› what odd or even or square or cube ‹signify›, and geometry ‹assumes› what irrational or inflexion or verging ‹signify›, but they demonstrate that they are through the common things and from what has been demonstrated. (Arist. *APo.* 1.10, 76^b6–11)

Science is not interested in proving the existence or the truth of per se attributes. Rather, it attempts to prove that these per se attributes belong to certain specific subjects, or *are* about them. So, Aristotle's formulation here seems to involve a converse predicative elliptical use of 'to be'. And this suggests that the formulation of the claim that one type of already existing knowledge is knowledge that certain things *are* covers not only the veridical and the existential use of 'to be', but also its predicative use (alluded to by ellipsis).[2]

This result is confirmed by *Posterior Analytics* 2, whose second chapter is dedicated to 'the things that we seek and those that on finding we know' (89^b36). Aristotle says:

T8 λέγω δὲ τὸ ὅτι ἔστιν ἐπὶ μέρους καὶ ἁπλῶς, ἐπὶ μέ- 90^a2
ρους μέν, ἆρ' ἐκλείπει ἡ σελήνη ἢ αὔξεται; εἰ γάρ ἐστι τὶ
ἢ μὴ ἔστι τί, ἐν τοῖς τοιούτοις ζητοῦμεν· ἁπλῶς δ', εἰ ἔστιν
ἢ μὴ σελήνη ἢ νύξ. 90^a5

I mean 'that it is' partially and *simpliciter* – partially: is the moon eclipsed? or is it increasing? (for in such cases we seek if it is something or is not something); *simpliciter*: if the moon or night is or is not. (Arist. *APo.* 2.2, 90^a2–5)

In passage T8 Aristotle informs his readers that the claim that we know about something that 'it is' covers both the claim that we know that it is so-and-so and the claim that it exists. Thus, the formulation of the claim that one type of already existing knowledge is knowledge that certain things *are* appears to cover most of the main uses of 'to be': veridical, existential, and predicative.

Knowing what certain things are: what use of 'to be' is left? Granted that one of the two types of already existing knowledge distinguished in T2, namely knowledge that certain things *are*, is formulated in a way that covers most of the main uses of 'to be', and therefore includes knowledge that certain things *are the case*, knowledge that certain things *exist*, and knowledge that certain things *are so-and-so*, one wonders what (logical) space could be left for the second type of already existing knowledge mentioned in T2, namely knowledge of what the thing said *is*, which Aristotle treats as equivalent to knowledge of what the thing said *signifies*. What use of 'to be' is left for knowledge of what the thing said *is*?

One possibility is that the second type of already existing knowledge should be included in the first as a subordinate case. This is perhaps the conclusion that one should draw, but I suspect that it is bound to cause some uneasiness: it would be strange to distinguish two cases the second of which turns out to be included in the first. It is therefore worthwhile to look for an alternative interpretation that avoids this unpalatable consequence.

Axioms, posits, suppositions, and definitions. An alternative interpretation may reasonably take its start from the observation that the second type of already existing knowledge, i.e. already existing knowledge of what the thing said is or (equivalently) of what it signifies, is knowledge that may be expressed in definitions. We should therefore consider what *Posterior Analytics* 1 says with regard to definitions (it seems prudent to restrict attention as far as possible to *Posterior Analytics* 1: *Posterior Analytics* 2 has a lot to say about definitions, but one gets the impression that the climate of *Posterior Analytics* 2 is different from that *Posterior Analytics* 1 and that one cannot systematically transfer the theses of the second book to the first).

---

[2] I am grateful to Marko Malink for pressing me on the interpretation of T6 and T7.

Not long after passage T2, in the middle of chapter 2 of *Posterior Analytics* 1, Aristotle discusses the types of principles involved in understanding, and some of these principles are definitions. Here is the relevant text (which, regrettably seems to be corrupt at two crucial points):

T9 ἀρχὴ δ' ἐστὶν ἀποδείξεως πρότασις ἄμεσος, 72a7
ἄμεσος δὲ ἧς μὴ ἔστιν ἄλλη προτέρα. πρότασις δ' ἐστὶν ἀντι-
φάσεως[3] τὸ ἕτερον μόριον, ἓν καθ' ἑνός, διαλεκτικὴ μὲν ἡ
ὁμοίως λαμβάνουσα ὁποτερονοῦν, ἀποδεικτικὴ δὲ ἡ ὡρισ- a10
μένως θάτερον, ὅτι ἀληθές. ἀπόφανσις δὲ ἀντιφάσεως ὁπο-
τερονοῦν μόριον, ἀντίφασις δὲ ἀντίθεσις ἧς οὐκ ἔστι μεταξὺ
καθ' αὑτήν, μόριον δ' ἀντιφάσεως τὸ μὲν τὶ κατὰ τινὸς κα-
τάφασις, τὸ δὲ τὶ ἀπὸ τινὸς ἀπόφασις. a14
Ἀμέσου δ' ἀρ- a14
χῆς συλλογιστικῆς θέσιν μὲν λέγω ἣν μὴ ἔστι δεῖξαι, μηδ' a15
ἀνάγκη ἔχειν τὸν μαθησόμενόν τι· ἣν δ' ἀνάγκη ἔχειν τὸν
ὁτιοῦν μαθησόμενον, ἀξίωμα· ἔστι γὰρ ἔνια τοιαῦτα· τοῦτο
γὰρ μάλιστ' ἐπὶ τοῖς τοιούτοις εἰώθαμεν ὄνομα λέγειν. θέσεως
δ' ἡ μὲν ὁποτερονοῦν τῶν μορίων τῆς ἀντιφάσεως λαμβά-
νουσα, οἷον λέγω τὸ εἶναί τι ἢ τὸ μὴ εἶναί τι, ὑπόθεσις, ἡ a20
δ' ἄνευ τούτου ὁρισμός. ὁ γὰρ ὁρισμὸς θέσις μέν ἐστι· τίθε-
ται γὰρ ὁ ἀριθμητικὸς μονάδα τὸ ἀδιαίρετον εἶναι κατὰ τὸ
ποσόν· ὑπόθεσις δ' οὐκ ἔστι· τὸ γὰρ τί ἐστι μονὰς καὶ τὸ εἶ-
ναι μονάδα οὐ ταὐτόν. 72a24

A principle of a demonstration is an immediate proposition, and immediate is one to which no other is prior. A proposition is one of the two parts of a contradiction, one thing about one thing. Dialectical is the one which indifferently assumes any ‹part› whatsoever, demonstrative the one which definitely ‹assumes› one of the two ‹parts› because it is true. A declaration is any part whatsoever of a contradiction, a contradiction is an opposition which, in virtue of itself, has nothing in the middle, a part of a contradiction is, on the one hand, something about something, an affirmation, on the other, something away from something, a negation.

An immediate syllogistic principle I call a posit if one cannot prove it but it is not necessary for anyone who is to learn anything to grasp it; and one which it is necessary for anyone who is to learn anything whatsoever to grasp, I call an axiom (for there are some such things); for we are accustomed to use this name especially for such things. A posit which assumes any part whatsoever of a contradiction – i.e.,[4] I mean, that either something is or something is not[5] – I call a supposition; one without this, a definition. For a definition is a posit (for the arithmetician posits that a unit is what is quantitatively indivisible). It is not a supposition (for what a unit is and that a unit is are not the same). (Arist. *APo.* 1.2, 72a7–24)

345

[3]The main MSS report 'ἀποφάνσεως', but such a reading is hard to make sense of because shortly later (at 72a11–12) an ἀπόφανσις is defined as a part of a contradictory pair, so that on the reading of the MSS Aristotle would be saying that a proposition is a part of a part of a contradictory pair. This motivates the emendation of 'ἀποφάνσεως' into 'ἀντιφάσεως' (note that Aristotle often speaks of the parts of an ἀντίφασις: cf. Int. 9, 19a37; 11, 20b23; *APr.* 1.1, 24a23–4; *APo.* 1.2, 72a11–12; *Metaph.* Γ8, 1012b13). A similar emendation is called for at 72a19: even if the reading of the MSS here does not generate an absurdity, the suggestion that a posit assumes 'any part whatsoever of an ἀπόφανσις' sounds wrong (one can imagine that the earlier corruption at 72a7–8 fostered this one).

[4]I give 'οἷον' an epexegetic value ('i.e.', 'namely'): cf. T. Waitz (ed. and comm.), *Aristotelis Organon Graece*, Leipzig 1844–6, I 280, 293; H. Bonitz, *Index Aristotelicus*, Berlin 1870, 502a1–23.

[5]One might translate 'τὸ εἶναί τι ἢ τὸ μὴ εἶναί τι' also by 'either it is something or it is not something'. The alternative translation would not make an important difference (because 'to be something' could cover both predication and existence), but I find the omission of the predicate

In the second half of passage T9 Aristotle puts forward a classification of immediate syllogistic principles:

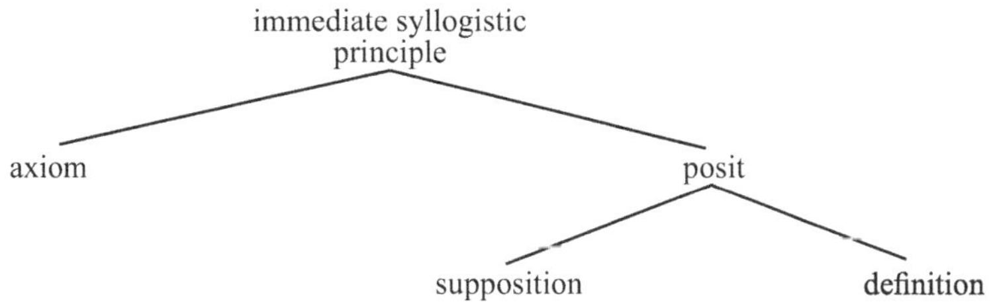

Axioms are completely general principles of a (roughly) logical character. They must be mastered in order to develop any discipline whatsoever. Posits are instead more specific principles: they must be mastered only by those who intend to develop certain specific disciplines (they need not be mastered by those who concentrate on other disciplines). There are two types of posits: suppositions and definitions. What distinguishes them is that a supposition 'assumes any part whatsoever of a contradiction—i.e., I mean, that something is or that something is not' (72^{a}19–20), whereas a definition is 'without this' (72^{a}21). At the end of T9 Aristotle tries to clarify the difference between suppositions and definitions by declaring that 'what a unit is and that a unit is are not the same' (72^{a}23–4). This remark seems to hark back to T2's distinction between two types of already existing knowledge (the example of the monad occurs there too, cf. 71^{a}15–16) and confirms that T9 is a passage that it is appropriate to examine in order to attain a better grasp of that distinction.

How can definitions fail to assume 'any part whatsoever of a contradiction'? According to passage T9, a supposition 'assumes any part whatsoever of a contradiction—i.e., [...] that either something is or something is not' (72^{a}19–20), while a definition is 'without this' (72^{a}21). Thus, a definition does not assume 'any part whatsoever of a contradiction—i.e., [...] that either something is or something is not'. How can this be the case? The problem is made even harder by the fact that in passage T9 seems to be internally incoherent. For, Aristotle appears to endorse the following claims: every definition is a posit (cf. 72^{a}21); every posit is an immediate syllogistic principle (cf. 72^{a}14–16); every principle of a demonstration is an immediate proposition (cf. 72^{a}7); every proposition is either of the parts of a contradiction (cf. 72^{a}8–9). It is difficult to refrain from assuming that by endorsing the claim that every posit is an immediate syllogistic principle Aristotle is committing himself to the claim that every posit is a principle of a demonstration. If this assumption is correct, then Aristotle is committed to the claim that every definition is a principle

(presupposed by my preferred translation) linguistically less harsh than the omission of the subject (presupposed by the alternative translation).

of a demonstration, therefore a proposition, and therefore either of the parts of a contradiction.

One might be tempted to think that in passage T9 'definition' means '*definiens*'[6]: since a *definiens* is a sub-sentential phrase, it is not a complete declarative sentence (cf. *Int.* 5, 17^{a}11–12), and therefore it is not one of the parts of a contradiction. Some support for such an interpretation might come from a passage in *Posterior Analytics* 1.10:

T10 Οἱ μὲν οὖν ὅροι οὐκ εἰσὶν ὑποθέσεις (οὐδὲν γὰρ εἶναι ἢ μὴ 76^{b}35
λέγεται), ἀλλ' ἐν ταῖς προτάσεσιν αἱ ὑποθέσεις, τοὺς δ'
ὅρους μόνον ξυνίεσθαι δεῖ· τοῦτο δ' οὐχ ὑπόθεσις (εἰ μὴ καὶ
τὸ ἀκούειν ὑπόθεσίν τις εἶναι φήσει), ἀλλ' ὅσων ὄντων τῷ
ἐκεῖνα εἶναι γίνεται τὸ συμπέρασμα. 76^{b}39

Terms are not suppositions (for nothing is said to be or not to be), but suppositions are among the propositions whereas one must simply grasp terms: and this is not a supposition (unless one were to say that hearing is a supposition). They [*sc.* suppositions] are among the things by whose being, in virtue of their being, the conclusion comes about. (Arist. *APo.* 1.10, 76^{b}35–9)

One might be tempted to translate the occurrences of 'ὅρος' in T10 by 'definition', in which case passage T10 would be saying something similar to passage T9 and would then support the interpretation of the latter passage according to which the definitions mentioned in it are simply *definientia*. Against this move one may observe that the occurrences of 'ὅρος' in T10 need not be rendered by 'definition': one may render them by 'term' (as I have done in my own translation of the passage). Against the rendering of the occurrences of 'ὅρος' in T10 by 'definition', and in favour of their rendering by 'term', note that what, properly speaking, needs to be grasped are neither *definientia* nor complete definitions, but *definienda*, which usually are terms (note the occurrence of 'ξυνιέναι' both in T2, at 71^{a}13, and in T10, at 76^{b}37: the role of a definition is precisely to enable one to ξυνιέναι the *definiendum*, and the definition would fail to do this if one were unable to ξυνιέναι its *definiens*). So, passage T10 provides no support for understanding the definitions of passage T9 as *definientia*.

Two considerations militate against the suggestion that the definitions mentioned in passage T9 are *definientia*, and therefore sub-sentential phrases. (1) In T9's explanation of why a definition should be called a 'posit', Aristotle says that 'the arithmetician posits that a unit is what is quantitatively indivisible' (72^{a}21–3). Granted that a posit is what a mathematician (or any other scientist) posits, Aristotle's explanation strongly suggests that a posit should be (not a sub-sentential phrase, e.g. 'what is quantitatively indivisible', but) a complete sentence (e.g. 'A unit is what is quantitatively indivisible'). (2) As I pointed out at the end of the penultimate paragraph, Aristotle may be plausibly be taken to be committed to the claim that every definition is a proposition. If this is right, then he cannot coherently claim that definitions are

[6] Cf. B. Landor, Definitions and Hypotheses in Posterior Analytics 72a19–25 and 76b35–77a4, *Phronesis*, 26 (1981) 308–18, at 311.

sub-sentential phrases, and in particular *definientia*.[7] In view of these objections, the idea that the definitions mentioned in T9 are *definientia* should be given up. We should instead assume that they are complete definitions. But then the problem of how a definition (a complete definitional sentence, e.g. 'A unit is what is quantitatively indivisible') can fail to assume 'any part whatsoever of a contradiction—i.e., [...] that either something is or something is not', returns as pressing as ever. How can Aristotle coherently claim this? Three solutions come to mind.

(1) Definitions are stipulations.[8] As such, they are instructions concerning the use of certain expressions and therefore they are not declarative sentences, so they do not assume 'any part whatsoever of a contradiction'. Moreover, they do not involve any of the main uses of 'to be' (veridical, existential, and predicative), or at least they do not involve any of these uses in connection with an indicative form of 'to be'. In support of this solution, note that as a matter of fact many of the definitions offered in mathematics are stipulative, and that mathematical definitions are those which Aristotle has primarily in mind in the passages in question. This solution has two drawbacks. First, it contributes nothing to solving the problem of how Aristotle could coherently be committed both to the claim that definitions do not assume 'any part whatsoever of a contradiction' and to the claim that every definition is a proposition and is therefore either of the parts of a contradiction. Secondly, stipulative definitions, being instructions, are not truth-evaluable. So, they cannot be true. As I pointed out at the end of the antepenultimate paragraph, Aristotle may be plausibly be taken to be committed to the claim that every definition is a principle of a demonstration. But, in the earlier part of chapter 2, Aristotle insists that 'the principles appropriate to what is demonstrated' (71^b23) must be true (cf. 71^b20–1, $^b25$–6). The only conceivable reply to this objection is to assume that Aristotle is allowing himself an extended use of 'true', whereby it applies also to stipulative definitions: 'true' should mean something like 'safe' or 'incontestable' (should one try to find a trace of this use of 'true' in the second half of *Metaphysics* Θ10?). However, such a reply feels ad hoc.

(2) Definitions do not describe what the world is like but spell out the terminological and conceptual apparatus whereby we speak and think about the world. For this reason, a definition does not assume 'any part whatsoever of a contradiction—i.e., [...] that either something is or something is not' (it is presupposed that the contradictory pairs which assume 'that either something is or something is not' consist of declarative sentences that speak of the world: the 'i.e.', which translates 'οἷον', explains the use of the present occurrence of the expression 'the parts of a contradiction' by narrowing it down to contradictory pairs that assume that something is or something is not) and does not display any of the main uses of 'to be' (it is presupposed that the main uses of 'to be', namely the veridical, existential, and predicative uses of the verb, are in this context restricted in such a way as to be concerned with describing what the world is like, and from this point of view the

[7]Cf. J. Barnes (trans. and comm.), *Aristotle, Posterior Analytics*, 2nd ed., Oxford 1993, 100.

[8]Cf. R. Robinson, *Plato's Earlier Dialectic*, Ithaca and New York 1941, 105.

differences among them are of secondary importance—for this reason they are all lumped together under the heading 'that it is'). However, being (accurate) accounts of the terminological and conceptual apparatus whereby we speak and think about the world, definitions can be qualified as 'true' in the usual sense of 'true' and are members of contradictory pairs: this is required by their being posits (cf. 72ª21), and therefore immediate syllogistic principles (cf. 72ª14–16), hence principles of demonstrations, hence immediate propositions (cf. 72ª7), where propositions are members of contradictory pairs (cf. 72ª8–9). The second solution has both historical and philosophical drawbacks. From a historical point of view, it is very speculative because there is no clear independent evidence for crediting Aristotle with a distinction like the one it draws. From a philosophical point of view, it is vague. For, the contrast between speaking of the world and speaking of the apparatus whereby we speak of the world is far from clear-cut. One might try to explain it by pointing to the distinction that modern philosophical logic draws between object-language and meta-language: the idea would then be that while suppositions are statements of (what we would call) the object-language, definitions are statements of (what we would call) the meta-language, and the reason why none of the main uses of 'to be' (namely the veridical, existential, and predicative uses of the verb) are involved in definitions is that it is being understood that the main uses of 'to be' are formulations that belong to (what we would call) the object-language. Despite Aristotle's remark that the type of already existing knowledge that he contrasts with knowing that certain things are consists in knowing what things signify (cf. 71ª14–16), a remark that fits well with the view that definitions are statements that belong to (what we would call) the meta-language, it must be avowed that the parallel with the distinction between object-language and meta-language feels far-fetched.

(3) There are no negative definitions: no definition is of the form 'A unit is not ...' (cf. *APo.* 2.4, 91ª26–9), and the first move of a definition is to place the kind indicated by the *definiendum* under its genus. For this reason, a definition does not assume 'any part whatsoever of a contradiction'. Aristotle is not denying that a definition is one of the parts of a contradiction: he is only denying that a definition can be any of the parts of a contradiction, and he is denying this because a definition cannot be the negative part of a contradiction. But why should definitions not display any of the main uses of 'to be'? Shouldn't they display the predicative use of 'to be'? As I pointed out earlier, these questions are all the more pressing because in the last sentence of passage T9 Aristotle connects the difference between suppositions and definitions with the difference between positing that something is and positing what something is. The most likely answer to these questions is that definitions involve a special way of understanding sentences, namely their 'definitional' reading as opposed to their 'ordinary' reading. While on its ordinary reading a sentence that is an instance of the schema 'σ is (a) π' is simply saying that the entity indicated by the expression that replaces 'σ' enjoys the attribute signified by the expression that replaces 'π', on its definitional reading a sentence that is an instance of the schema 'σ is (a) π' uses the expression that replaces 'π' to offer a complete description of the nature of the

entity (typically a kind) indicated by the expression that replaces 'σ'. For instance, the sentence 'Man is a biped terrestrial animal' is true on its definitional reading but false on its ordinary reading: it is true on its definitional reading because 'biped terrestrial animal' is a complete description of the nature of the entity indicated by 'man', namely of the kind man; it is false on its ordinary reading because the entity indicated by 'man', namely the kind man, does not enjoy the attribute signified by 'biped terrestrial animal' (kinds are not animals and do not have feet). Definitions are typically supposed to be understood in accordance with the definitional reading. Aristotle is lumping together the veridical, existential, and predicative uses of 'to be' in so far as they can all be brought under the heading of the ordinary reading of sentences involving them (because in all of them the entity indicated by the subject-expression is said to enjoy a certain characteristic, i.e. to be the case, or to exist, or to be so-and-so). This solution has some historical plausibility because there are reasons for thinking that the distinction between the ordinary and the definitional readings of sentences played an important role in Plato's *Sophist*,[9] which Aristotle knows well.

An objection might be raised against the third solution: one passage of the *Posterior Analytics* (2.4, 91^a26–32) seems to presuppose that definitions are universal affirmative propositions, and three other passages (1.8, 75^b30–2; 2.3, 90^b24; 17, 99^a21–2) appear to indicate that definitions can be used as premisses in demonstrations.[10] Doesn't this evidence tell against the view that definitions are full descriptions of the nature of kinds? Now, it is not clear how much weight may be given to the first passage, which belongs to a dialectical context where the idea that definitions can be demonstrated is put to the test. After all, Aristotle could simply be saying that *if* definitions could be demonstrated, *then* they would have to be universal affirmative propositions. As for the three passages where definitions appear to be treated as premisses of demonstrations, one of them (2.17, 99^a21–2) merely affirms that demonstrations come about 'through definition' (99^a22) whereas the other two (1.8, 75^b30–2 and 2.3, 90^b24) claim that definitions are principles of demonstrations. This claim should not come as a surprise: after all, passage T9 is explicitly classifying definitions as principles of demonstrations. And, of course, the third solution is not committed to denying that definitions play the role of principles of demonstrations. The crucial point of the third solution is that the foundation of the definitions' role as principles is their being full descriptions of the nature of kinds, and definitions can play such a role by giving rise to, perhaps even by taking the form of, universal affirmative propositions that feature among the premisses of demonstrative syllogisms: even if 'Every unit is a quantitatively indivisible entity', or 'Units are all and only the quantitatively indivisible entities', figures among the premisses of a demonstration,

[9]I argued for an interpretation of the *Sophist* that credits Plato with a distinction between the ordinary and the definitional reading of predicative sentences in my *Plato's Account of Falsehood: A Study of the* Sophist, Cambridge 2012, 122–36.

[10]Cf. M. Mignucci (trans. and comm.), *Aristotele, Analitici secondi*, Rome and Bari 2007, 157.

the definition that grounds it, or is perhaps even expressed in it, namely the definition 'A unit is what is quantitatively indivisible' (72^{a}22–3), remains a complete description of the nature of the kind unit, and cannot be described as saying that the kind unity enjoys certain attributes. So, despite the objections that may be levelled at it, the third solution provides a likely reply to the difficulties raised by the first two chapters of the *Posterior Analytics*.

Trust in scientific principles. Immediately after offering his classification of kinds of scientific principles, in the last part of *Posterior Analytics* 1.2, at lines 72^{a}25–^{b}4, Aristotle addresses the issue of the degree of conviction of scientific principles, which he construes as principles of demonstration. He argues that demonstration is the means through which we become convinced of a proposition, which is the demonstration's conclusion, and that the premisses must be the causes of the conviction generated by the demonstration. He also assumes a causal principle: if x is the cause for y's being F, then x must be F to a higher degree than y. On the basis of these considerations he infers that the principles of a demonstration must be propositions we are more convinced of than its conclusion. In fact, he goes even further: he asserts that the conviction one has with respect to scientific principles must be the greatest possible in that 'there must be no other thing more convincing to him [*sc.* the demonstrator] or more familiar among the opposites of the principles on which a syllogism of the contrary error may depend—if anyone who understands *simpliciter* must be unpersuadable' (72^{b}1–4). Aristotle also alludes (at 72^{a}33–4) to the possibility of being 'better disposed towards' certain propositions than by knowing them.

One wonders how these considerations about degrees of conviction about scientific principles tie in with the immediately preceding classification of them. In the case of axioms, it is easy to see why Aristotle may feel in a position to regard them as maximally convincing: after all, in his discussion of logical principles in the *Metaphysics* he describes the most eminent of them, namely the Principle of Non-Contradiction, as the most certain of all (cf. *Metaph.* Γ3, 1005^{b}8–23). As for definitions, the allusion to being 'better disposed towards' certain propositions than by knowing them might be a hint that there is a special faculty, namely intellect (νοῦς), whose function is to grasp definitions, and in general to put the thinker in a mental state whose proper formulations involve sentences that are to be understood in accordance with their definitional reading. The intellect must then be a capacity to 'look into' the nature of the kinds it is concerned with. If this is Aristotle's position, we must avow that it is more the statement of a desideratum than a proper proof, or even an argument, to the effect that scientific principles of this sort enjoy the elevated epistemic status that Aristotle ascribes to them. We are obliged to draw a similar conclusion with regard to the other type of posits distinguished by Aristotle, namely suppositions: Aristotle requires them to be most convincing and trustworthy, but he gives no indication of how this trustworthiness is to be achieved. In this respect, the *Posterior Analytics* are more a normative description of what a science should be like than a recipe about how to construct a proper science.

References

Barnes J (trans. and comm.) (1993) Aristotle, Posterior Analytics, 2nd edn. Clarendon Press, Oxford
Bonitz H (1870) Index Aristotelicus. Reimer, Berlin
Crivelli P (2012) Plato's Account of Falsehood: A Study of the Sophist. Cambridge University Press, Cambridge
Landor B (1981) Definitions and Hypotheses in Posterior Analytics 72a19–25 and 76b35–77a4. Phronesis 26(1981):308–318
Mignucci M, (trans. and comm.), (2007) Aristotele. Analitici secondi, Laterza, Rome and Bari
Robinson R (1941) Plato's Earlier Dialectic. Cornell University Press, Ithaca and New York
Waitz T (ed. and comm.) (1844–1846) Aristotelis Organon Graece. Hahn, Leipzig

From Antiquity to the Middle Ages. Philosophy, Truth, and Demonstration

Cristina D'Ancona

Abstract A common saying of medieval Aristotelianism, both in Arabic and in Latin, labels Aristotle the "First Teacher". This label implies not only that Aristotle established the system of sciences, but also that he created the demonstrative method. Against the background of the contemporary debates about demonstration and inquiry on natural phenomena in Aristotle, this article contends that the medieval pattern of the "First Teacher" is grounded in late Antiquity, and inherits from the vision of science elaborated first by Alexander of Aphrodisias, and then in the Neoplatonic school of Alexandria.

Keywords Aristotle · Demonstration · Order of the sciences · Commentaries · Alexander of Aphrodisias · John Philoponus · Al-Kindī · Averroes · Late antiquity · Arabic reception · Latin reception

[1] Lloyd (1990), 371: "The question of the relationship between Aristotle's theory of science set out in the *Posterior Analytics* and his actual practice in the physical treatises is, you will feel, a hoary old chestnut indeed. Faced with a series of apparent discrepancies between the two, starting with the notorious mismatch between the attention devoted to the theory of the syllogism in the *Organon* and evident lack of good-looking actual examples of syllogisms in the physical works, what are we to say?" I owe to one of the referees of this article the remark that Lloyd's tenet is best accounted for against the background of his more general refusal to see Aristotle as engaged with the question of how to reach a systematic vision covering all the scientific fields. The same referee also points to Gotthelf's reply to Lloyd in Gotthelf (2012), 185: both remarsk shed light on the issue at hand.

My most sincere thanks go to Concetta Luna. Her reading of a first draft of this paper saved me from many errors. I am also indebted to the anonymous referees for this volume, whose helpful criticisms are acknoweldged in the footnotes below and to whom I express here my heartfelt thanks. All the remaining shortcomings are mine.

C. D'Ancona (✉)
University of Pisa, Department of Civilizations and Forms of Knowledge, via dei Mille 19, 56126 Pisa, Italy
e-mail: cristina.dancona@unipi.it

A. Fabris (ed.), *Trust*, Studies in Applied Philosophy, Epistemology and Rational Ethics 54, https://doi.org/10.1007/978-3-030-44018-3_2

1 The Old Chestnut

Scholars working on Aristotle are familiar with a debate that was only apparently dismissed by G.E.R. Lloyd[1] as an old chestnut: that on the tension between the theory of demonstrative science and the procedure Aristotle himself adopts in studying natural phenomena. In reality, the debate was reactivated by Lloyd's paper. Accounting for empirical observations is indeed the aim of the science of nature[2]; but whether or not empirical observations and the accounts these give rise to meet the criteria for a body of knowledge to be demonstrative science in Aristotelian terms was the subject of a prolonged controversy since at least 1961, when a well-known essay by G.E.L. Owen entitled '*Τιθέναι τὰ φαινόμενα*'[3] pointed to the aforementioned tension. Owen remarked that while in the *Prior* and *Posterior Analytics* Aristotle establishes observed facts as the basis for elaborating demonstrations whose deductive structure counts as the criterion of science, in the *Physics* he does not himself follow the procedure he had recommended. Other Aristotelian scholars were dealing with this problem in the same years, most prominently W. Wieland[4]; further voices joined the debate,[5] until a solution to the 'classical problem' was suggested by J. Barnes in 1969: what Aristotle describes in the *Posterior Analytics* is a method of instruction rather than a procedure of scientific discovery.[6] Later on, this approach was referred to as the "Teaching Thesis".

Despite the fact that Barnes' solution has not met with unqualified scholarly consensus,[7] the "Teaching Thesis" represents a sort of vantage point for the historian of philosophy. It goes without saying that what inspired the late Antique and medieval commentators of Aristotle was a remarkably different vision. However—and without

[2]E.g. in *Part. An.* I 1, 639 b 8–10. See also *An. Post.* I 13, 78 b 39—79 a 5; *De Cael.* II 13, 293 a 23–30; III 7, 306 a 5–17; *Metaph.* XII 8, 1073 b 32–38.

[3]Owen (1961).

[4]Wieland (1960–61), English version (1975).

[5]A further question that cannot be dealt with here is the relationship this debate has with that on the status of demonstrative science versus dialectic: cf. for instance Hamlyn (1990).

[6]Barnes (1969, 1975), 77 and 85 (of the 1975 revised version): "The theory of demonstrative science was never meant to guide or formalise scientific research (…) it does not describe how scientists do, or ought to, *acquire* knowledge. (…) In developing the theory of demonstration and in constructing his notion of demonstrative science, Aristotle was not telling the scientist how to conduct his research: he was giving the pedagogue advice on the most efficient and economic method of bettering his charges. The theory of demonstration offers a formal account of how an achieved body of knowledge should be presented and taught". As the earliest advocate of this solution, Barnes (fn. 70) quotes Grote (1880). Barnes' interpretation is labelled "Teaching Thesis" by Wians (1989).

[7]Burnyeat (1981) raises doubts about the existence of a 'pedagogic' approach in the *Posterior Analytics*, whose aim is rather to generate complete understanding of demonstrative truths about nature: only such a complete understanding meets the criteria to be a science, and complete understanding is not within the reach of a disciple. Barnes (1981) partly reformulates his position, and other critical remarks are advanced by Wians (1989), also against the revised version. One of the referees of this paper remarks that contemporary Aristotelian scholarship the "Teaching Thesis" is not seen as a viable solution to the "classical problem" mentioned above.

delving into the question whether or not the "Teaching Thesis" offers the way to mitigate a perceived inconsistency that seems to undermine Aristotle's project of science—the idea that Aristotle *teaches* philosophy as a structured body of knowledge stands, so to speak, in the background of the Aristotelian commentary tradition: much of what I will say in Sect. 2 counts as a footnote to this claim. Aware as I am that this thumbnail sketch is unsatisfactory,[8] I will nevertheless follow it as a red herring to outline, in Sect. 2, the late antique context that paved the way for Aristotle to play in the Middle Ages the role of the First Teacher—*al-mu 'allim al-awwal, il maestro di color che sanno*—as will be described in Sect. 3.

2 The "Teaching Thesis" and Aristotelianism from the Imperial Age to Late Antiquity

It is against the backdrop of the end of Hellenism and the beginning of the Imperial age that philosophy, especially Aristotelian, saw the transformation into what can be labelled an era of exegesis.[9] The rediscovery of Aristotle's (and Theophrastus') library, described by Strabo and Plutarch and commented upon time and again,[10] was of pivotal importance, even taking into account the caveats that have been advanced on the direct relationship between the edition of Aristotle's unearthed works and the rebirth of interest in his philosophy after a long decline.[11] A point which is especially

[8]Significant contributions to the solution of the riddle include Bolton (1987 and 1991), Ferejohn (1990), Berti (1991), McKirahan (1992), Barnes (1993), Lennox (2001), Falcon (2005), Ebrey (2015), Falcon (2017).

[9]Hadot (1987a, b), 14: "A partir du 1er siècle av. J.C., avec la ruine de la plupart des institutions philosophiques d'Athènes, provoquée par les dévastations de Sylla, avec la formation de nombreuses institutions philosophiques dans l'ensemble du Bassin méditerranéen, une seconde phase de l'histoire de la philosophie postsocratique se développe. Les quatre tendances doctrinales fondamentales subsistent, mais elles ne sont plus supportées par l'institution athénienne créée par les fondateurs. Pour affirmer leur fidélité au fondateur, le quatre écoles philosophiques, répandues dans différentes villes d'Orient et d'Occident, ne peuvent plus s'appuyer sur l'institution qu'il a créée, ni sur la tradition orale intérieure à l'école, mais uniquement sur les textes du fondateur. Les cours de philosophie consisteront donc avant tout dans des commentaires de ce texte". Cf. also Donini (1994) and Sedley (1996). The Lyceum ceased to exist as an institution after Sulla's conquest of Athens: cf. Lynch (1972), 192–207; as for the Platonic school, it is well known that the Platonists disagreed on the question whether or not with Arcesilaus of Pytane (d. 241 b.C.) the Academy continued to be the same school that had been founded by Plato; one thing however is sure, it ceased to exist as an institution after Sulla's conquest.

[10]Düring (1957); Moraux (1973), 3–94; Gottschalk (1987), 1083–1097, Barnes (1996), Hatzimichali (2016). My thanks go to one of the referees for directing me also to Primavesi (2007).

[11]Sandbach (1985) argued that Aristotle was little known during the Hellenistic age; critical remarks on this position were advanced by Hahm (1991); see now Falcon (2016), and Bénatouïl (2016). On the rediscovery of the corpus, the sources that record the facts, and the many problems related to this, cf. above n. 10.

germane to the present discussion is that when the corpus was edited by Andronicus of Rhodes, in the first century b.C.,[12] this was done according to a systematic structure that implies a precise propaedeutics. Logic, in this systematic edition, is followed by physics, which includes the science of the soul; physics is followed by metaphysics, and metaphysics by ethics and politics. To what extent does this systematic structure depend upon Aristotle's himself, though of course in association with the post-Aristotelian emergence of the topic of the 'parts of philosophy',[13] is not easy to determine. One thing, however, seems to be clear: much in the (admittedly post-Aristotelian) organization of the corpus depends upon Aristotle's own explicit account of what we might label the forefather of the Medieval *subalternatio* of the sciences, were it not for the caveats that such a label can suscitate.[14] What is surely present in the *Posterior Analytics*[15] is the idea that in each primary field the departmental sciences take their principles from an overarching science. Particularly relevant to the issue at hand here is the fact that Aristotle provides also a presentation of his own treatises on natural sciences, and one that seems to be inspired by such criteria. This he does at the beginning of the *Meteorologica*.

> We have already dealt with the first causes of nature and with all natural motion; we have dealt also with the ordered movements of the stars in the heavens, and with the number, kinds and mutual transformation of the four elements, and growth and decay in general. It remains to consider a subdivision of the present inquiry (*μέρος τῆς μεθόδου ταύτης*) which all our predecessors have called meteorology. Its province is everything which happens naturally, but with a regularity less than that of the primary elements, and which takes place in the region which borders most nearly on the movement of the stars (*Meteor.* I 1, 338 a 20—b 22, trans. Lee, p. 5).

Recent scholarship has raised also apropos this part of Aristotle's natural science the puzzles of the "Old Chestnut" that had been previously raised mostly about the biological works, and whose solution (if any) cannot retain us here.[16] But Alexander

---

[12]The role of Andronicus' edition in the rebirth of Aristotelianism has been challenged by Barnes (1996), 66, as follows: "Nothing suggests that the 'Roman edition', done by Andronicus of Rhodes, revolutionized Aristotelian studies. His text of Aristotle left little mark on posterity. His work as orderer and arranger of the treatises was not epoch-making. (…) There is no reason to think that the Peripatetic renascence was any more dependent on books; and there is no reason to think that Andronicus played midwife at the rebirth".

[13]Hadot (1979).

[14]I owe to one of the referees of this article the remark that we should be wary of speaking of *subalternatio*: subordination as discussed in the *Posterior Analytics* does not apply primarily to the relationship of a science to higher or lower ones; rather, it applies to the relationship between a science and its subfields. This is surely true, especially in consideration of Aristotle's rejection of a unique overarching science, that results from the prohibition of kind-crossing explanations; however, there is a tension (not to be explored here) between this tenet and the theory of the "most architectonic science" of *Metaphysics* A 2 (see below note 15, and cf. McKirahan 1978).

[15]*An. Po.* I 13, 78 b 35–39. For first philosophy, or metaphysics, as the overarching science that encompasses every other one because it investigates the first causes of the whole reality, see *Metaph.* A 2, 982 a 14—b 12.

[16]R. Chiaradonna offered a useful discussion of this question and the various responses to it at the meeting of the Centre "Greco, arabo, latino" of the universities of Pisa—Padua—EPHE Paris held in

of Aphrodisias has no doubt that Aristotle engages in a description of the entire physical science according to the order (τάξις) in which this science has to be taught. The τάξις is dictated by the principles of *subalternatio*: first come the general principia, then the departmental sciences develop within and below the general principia. Here is what Alexander says in the opening pages of his commentary.

> At the beginning of his *Meteorology*, the first thing that he does is to remind us what he had said in the treatise (πραγματεία) on the *Physics*, and at one and the same time he also presents us with the order of the entire physical doctrine (θεωρία), both of that which has been expounded before and of that which will be accounted for after this one. First of all he recalls the course (ἀκρόασις) entitled *Physics*, as it is the first treatise and through it he recalls the entire physical doctrine. It is in fact in this treatise that he has dealt with the first principles and causes of nature, as well as with all the things that cooperate with natural movement: these are the items he recalls first here in general terms. Second he recalls the treatise *On Heavens*, for it is in this treatise that he has dealt "with the ordered movements of the stars in the heavens, and with the number, kinds and mutual transformation of the four elements". (…) Third he recalls *On Generation and corruption*, for it is in this treatise that he has dealt with the transformation of the first bodies one into another, a topic that had been explored also in the third book of *On Heavens*: but here he drives to completion his treatment of this transformation, explaining that there is such a transformation, what it is, and in what respect it differs from the other transformations. He assigns the fourth place to the present treatise, that he also entitles *Meteorology* because, as he says, his predecessors named it 'meteorology'. He labelled 'method' both the examination of nature and the science of it. (…) He also adds the items that, coming after the treatise on meteorology, still belong to the treatise on nature. In fact, he says: "and about animals and plants, both in general and in particular" (339 a 8). Indeed, he can include in the study of animals also *On the Soul*, and again *Sense and sensibilia*, *On Memory*, *On Sleep*, *On Divination in Sleep*, and again *On Youth and Old Age*, *On Length and Shortness of Life*, and all the other works he wrote that refer to animals: of these, some include the general consideration of all animals, like the *Inquiries on Animals*, *Generation of Animals*, *Parts of Animals*, *Movement of Animals*, *Progression of Animals*; others include only specific treatments, like *On Memory*, *On Sleep*, and *On Divination in Sleep*: what is dealt with in such treatises regards for the most part only mankind. Speaking of the primary "inquiry" he sets for himself to expound the entire physical doctrine: this is what he had set for himself as a goal. He said "according to the mode already said" (339 a 6–7) because in what comes before this treatise he did not limit himself to mere description: rather, in addition to inquire and expound the specific causes of each of the mentioned items, he intended also to present the doctrine accompanied by scientific demonstration. In the same vein he wants also to deal with the topic at hand here (Alexander of Aphrodisias, *On Aristotle's Meteorology*, pp. 1.5–4.11 Hayduck, my trans.).

It can well be the case that Alexander is simply following the order of the treatises that, in the edited corpus, are placed between the *Physics* and the *Metaphysics*—an order that may simply reflect and amplify Aristotle's own outlines, provided at the beginning of the *Meteorology* (quoted in part above) as well as elsewhere, chiefly in *On Sense and sensibilia* 436 a 1–5, where he says that the new inquiry takes for granted the conclusions already reached in the *De Anima*.[17] Note, however, Aristotle's

---

Pisa, November 26–28, 2019, devoted to the tradition of Aristotle's *Meteorologica* in late Antiquity and in the Arab and Latin worlds. On the issue of the biological works cf. Gotthelf and Lennox (1987) Devereux—Pellegrin (1990); on the *Meteorology*, see Wilson (2013), Scharle (2015).

[17]Needless to say that the passage of the *Meteorologica* does not imply an order of composition, as if Aristotle had written first the *Physics*, then the *De Caelo*, then again the *De Gen. Corr.*, and

stance in the account of Alexander: the overview of all the parts of his own φυσικὴ θεωρία—those which have already been written, and those that he plans to write after the *Meteorology*—allows him to organise the body of knowledge into sets containing subsets, until all the departmental sciences are included. Not only, but the mode of exposition, far from being merely descriptive, ψιλὴ ἱστορία, is scientific, which means demonstrative: μετὰ ἀποδείξεως. Admittedly, this is not the "Teaching Thesis": Alexander sees no inconsistency whatsoever between the criteria for science in the *Posterior Analytics* and the inquiry about nature; hence, he does not try to solve it. On the other hand, if a body of knowledge forms a science with all its subsets this owes much, in Alexander's eyes, to the fact that such a body of knowledge can be taught in good order, starting from general principles and descending until all particular phaenomena are accounted for, and μετὰ ἀποδείξεως.

On this ground, it comes as no surprise that centuries later, when the Neoplatonists of the school of Ammonius Hermeiou started lecturing on Aristotle in Alexandria, this model was fully endorsed within the context of a curricular structure of teaching—a structure that after the studies of Ilsetraut Hadot has no need to be described again.[18] Among the typical features of the Alexandrian curriculum there are the 'points' of the so-called 'Prolegomena' to the study of philosophy in general, and Aristotelian philosophy in particular. These 'points' to be dealt with when lecturing on a curricular text—which in Alexandria was chiefly an Aristotelian work—include the purpose of a given treatise, its usefulness, authenticity, place in the order of reading, part of the philosophy it belongs to, subdivision into chapters, and mode of teaching. The way in which John Philoponus comments upon the beginning of the *Meteorologica* shows the influence of Alexander, combined with the curricular points typical of the school (remember that Philoponus was the pupil of Ammonius Hermeiou and the editor of some of his lecture courses).

> We have already said before that philosophy can on good grounds be divided into two parts, the practical and the theoretical, because the faculties of our soul are also two, a vital one and a theoretical one, to which philosophy strives to bring order and perfection: to the one through virtue, to the other through knowledge of reality, since each thing is brought to perfection by attaining its own end for which it came into being. Hence philosophy is said to be assimilation to God as far as possible; for of divine actions, too, some are directed to the knowing of reality, others to creating and providing for it. Now into each of our two activities, i.e. theory and practice, some kind of corruption insinuates itself: in the case of theory this is falsehood; in the case of practice, it is vice. Therefore philosophy needed a tool for division: in its theoretical part one that distinguishes falsehood from truth, in its practical part, one that distinguishes vice from virtue (…). This [tool] is the so-called system of logic, which Aristotle has taught us in several books, and from it we were trained in the science of demonstration. Since the theoretical part of philosophy is divided into three parts, physics, mathematics and theology, the philosopher has treated physics more fully than the others, as being on our own level and more akin. I have elsewhere explained its division in detail; the present work also belongs to it. As we prepare to deal with this work, we must of necessity first expound its purpose. (…) The usefulness of the work is obvious (…). Nor is its

finally the *Meteorologica*: the point is discussed by Burnyeat (2004). On the relationship between the *Parva naturalia* and the *De Anima* cf. Sassi (2014), with discussion of previous literature; cf. also Bydén and Radovic (2018).

[18] Hadot (1987a, b, 1991).

authenticity subject to doubt. The preface will instruct us precisely on this point: explaining its place in the series and its position with respect to his other writings, Aristotle lists his works on physics and shows that this one logically follows upon those and forms a sequence with the ones that come after it. From this, both the authenticity and [its place in] the order [of reading] become manifest. The first place among all his works on physics is allotted to what is more specifically named the *Course on Physics*, second after it comes *On Heaven*, third *On Coming to be and Perishing*, and after that the present work; we shall state the reasons for this when explaining the text. The title is *Meteorology* because the things that he examines take place in the region between earth and heaven (pp. 1.4–2.22 Hayduck, trans. Kupreeva 2011, pp. 29–30).

In Philoponus' account, Aristotle is committed to granting the truth of the science of nature as a whole. The standard of knowledge is demonstrative method,[19] but this now implies teaching (and learning) scientific truth about nature and its separate causes according to the complete order of the disciplines, bringing from logic to physics, and from physics to metaphysics. This full-fledged version of the systematisation of the Aristotelian corpus presupposes an even broader frame, that exceeds the limits of Andronicus' edition: philosophy equals the Platonic assimilation of the human intellect to the divine mind, and the latter can be pursued under Aristotle's guidance precisely because he has provided us with the foundations of knoweldge—logic—and the science of nature, treating physics "more fully than the others". Then, mathematics will enable us to deal with separate truths; then again, theology (in the Neoplatonic sense) will form the pinnacle of the whole system of knowledge. Aristotle himself would totally agree with this, because—Philoponus' allusion to Book I of the *Physics* is telling—one should start from what is more akin to us, if one wants to proceed to what is more intelligible in itself, though remaining obscure to uneducated minds. The Platonic model of the exit and ascent from the Cave imperceptibly comes to the fore, and the order of the Aristotelian corpus is encapsulated between the initial claim that the goal of philosophy is the ὁμοίωσις θεῷ κατὰ τὸ δυνατόν and the final allusion to the intermediate level of the mathematical sciences paving the way to the study of divine realities, i.e. the intelligible separate principles of the visible world. Against this backdrop, it comes as no surprise that in Arabic philosophy Aristotle could have been accepted as the author of the pseudo-*Theology* (in fact, Plotinus) and of the *Liber de causis* (in fact, Proclus).

3 A Late Antique Pattern for 'Medieval' Philosophy in Arabic and Latin

The Alexandrian 'Introductions to philosophy' began to circulate in translation from the formative period of Arabic-Islamic philosophy onwards.[20] Starting from the ninth century, Arab readership got in touch with the Prolegomena presenting not

[19]On the late Antique interpretations of Aristotle's doctrine of demonstrative science see De Haas (2002), De Haas et al. (2010).

[20]Hein (1985).

only the systematic structure outlined above, but also the list of the definitions of philosophy, that includes that of 'assimilation to God'.[21] Even more importantly for the present purpose, the transformation of the Aristotelian corpus into a pyramidal structure culminating in the treatment of theology, which is only aired in Philoponus' passage, is presented in as many words in a treatise on the Aristotelian works by the first philosopher of the Arabic-speaking world, al-Kindī (d. circa 870). In his *Letter on the Quantity of Aristotle's Books and What is Required for the Attainment of Philosophy* al-Kindī writes:

> Aristotle's books have an order and an arrangement which the student must follow when perusing them one after another, so that he may thereby become a philosopher. After the propaedeutics, they are of four kinds. The first of four is logic. The second is physics. The third deals with what has no need for nature, subsists in itself, and does not require bodies; it exists together with bodies and is connected to them in some way. The fourth deals with what does not require bodies and is completely inconnected to them. There are eight books on logic. The first is called *Categories* and deals wit terms, I mean subject and predicate. (…) The second book on logic is called *Peri Hermeneias*, which means *On Interpretation*, that is, 'interpretation' of what is said through terms and their conjunction (…). The third book on logic is called *Analytics*, meaning 'the end from the beginning'. The fourth work on logic is called *Second Analytics*, but it is more properly named *Apodeictics*, which means proof. The fifth book on logic is called *Topics*, which means 'places', here meaning places in speech. The sixth book is called *Sophistics*, and its meaning is related to the 'sophists', which means 'wise men'. The seventh book on logic is called *Rhetoric*, which means [the science of eloquence]. The eighth book on logic is called *Poetics*, which means [the science of] poetry. This then is the quantity of the eight books on logic. As for the quantity of the ordered books on physics which one absolutely needs if one intends to achieve knowledge of natural things in order to become a philosopher in this regard, they are seven. The first is the book which is called *Lecture on Physics*. The second is called *On the Heavens*. The third is called *On Generation and Corruption*. The fourth is called *On Athmospheric and Terrestrial Phenomena*, which is also known as *The Things on High*, and as *Treatise on Extremes*, that is, things that are opposed to one another. The fifth is his book called *On Minerals*, the sixth his book called *On Plants*, and the seventh his book called *On Animals*. There are four books on things that have no need for bodies for their persistence and subsistence, but exist together with bodies. The first is his book called *On the Soul*. The second is his book called *On Sense-Perception and Sense-Objects*. The third is his book called *On Sleep and Waking*. The fourth is his book called *On the Length and Brevity of Life*. There is one book on things that have no need for bodies and are unconnected to bodies: his books designated the *Metaphysics*. After all these, he composed books which, although they belong to the kind of writings connected with the soul, become perfectly clear and that can be discussed in a plain manner only after one has dealt with things which do not need bodies at all. They [*sc.* the books on ethics] are the fruit of knowing these books [*sc.* on metaphysical issues], which produce the benefit derived from comprehending them (*Letter on the Quantity of Aristotle's Books*, pp. 364.10–368.17 Abū Rīda, trans. Adamson—Pormann, pp. 283–284).

The transformation of the method of inquiry into a method of instruction that some detect as Avicenna's own discovery is, in the last resort, nothing more than an echo of the late antique exegesis. It is this stance that inspires to al-Kindī even

[21] Hein (1985), 86–130.

the detail of classifying the *Rhetoric* and *Poetics* as a part of logic.[22] To this, al-Kindī adds a hierarchy of the scientific fields that echoes—with qualification—the passage by Philoponus mentioned above. The difference is that in the place of the mathematics, indicated by Philoponus as the discipline that steers a middle course between corporeal and incorporeal realities, in al-Kindī we find the science of the soul with its subsets.

Also in Averroes' Prologue to his Long Commentary on the *Physics*[23] the Alexandrian echo is easy to detect.[24] Averroes (d. 1198) sets for himself the task of commenting upon the *Physics* in the following way:

> My purpose in this discourse is that of commenting upon Aristotle's book on physical doctrine (...). Therefore I will first follow the custom of the commentators, who at the beginning of a book list eight main points, namely the purpose of the book, its usefulness, its place in the reading order, the title of the book, and the name of the author. (...) Thus in my opinion the purpose of natural science taken as a whole, of which this book is part and parcel, consists in knowing the causes of visible things as well as the causes of the accidental properties present in them, because the subject-matter of this science is represented by the visible things that are changeable in and by themselves, since they have in themselves the cause of motion and rest. Now, the purpose of this book consists in knowing about these things the causes that are common to all natural things, i.e. matter, form, the final cause, and the efficient cause; in addition, its purpose is to know the first causes, to the extent in which it is possible to know them within the province of this science, namely first matter and the first mover. On the contrary, the first formal cause and the first final cause fall under the consideration of the metaphysician. (...) The place of the reading order of this book consists in that this book is the first among those devoted by Aristotle to physical doctrine, because the natural things are subdivided into different natures, and notwithstanding the fact that they are diverse according to the diversity of their specific features, they have nonetheless aspects which are common to them all. This is why this science follows two lines of argument: the first one is the consideration of the aspects which are common to all natural things; the second one is the consideration of the aspects that are specific to each kind of the different natures. This second line is subdivided according to the number of the different kinds, and it is for this reason that in this art various books have been included, exactly as it happens in logic and in all other theoretical arts. (...) This is why here too it is necessary to begin by a part common to all, exactly as the *Prior Analytics* and *On Interpretation* are placed before the *Posterior Analytics*, the dialectic, and so on. The name of the author is Aristotle, son of Nicomachus. The wisest among the Greeks, he has composed other books in this art, in the fields of logic as well as of metaphysics. It was he who discovered and brought to completion these three arts. He discovered them, because all that had been written by the Ancients in this science is not worth being included as a part of it, not even as a source of doubts, least of all as something that can serve as a principle: it is he who brought it to perfection, because nobody among those who came later, up to the present age—which means one thousand and five hundred years—was able to add even a bit, or to detect in his words even one single mistake, even the

[22]On the so-called "enlarged Organon" of late Antiquity and its Syriac and Arabic fortune cf. Hugonnard-Roche (2007).

[23]This commentary is lost in Arabic and has come down to us in Latin and Hebrew translations. The passage above is translated from the Latin version. We owe to Harvey (1985) an English translation of the Hebrew version of the Prologue.

[24]Lack of space forbids the treatment of the connected topic of the so-called "enumeration of the sciences" in al-Fārābī, his predecessors, and the philosophers inspired by him—chiefly Avicenna. This "enumeration of the sciences" was one of the main conduits of the late Antique curriculum to the Latin Middle Ages, via Dominicus Gundissalinus.

> lesser. That there has been such a force in one single human person is a miracle, something extraordinary. Such a status, if it occurs in one individual man, has to be acknowledged as divine, rather than as human (*Long Commentary on the Physics*, 1r–v, 3 r–v, 4v—5r, my trans.).

Lack of space forbids altogether a treatment of this text as it deserves.[25] In particular, a comparison with Philoponus' commentary on the *Physics* would show that Averroes is following in his footsteps, taking from his Alexandrian predecessor the main points of the curricular "Introductions to philosophy": purpose, place in the reading order, authenticity.

The final *laudatio* may at first glance generate the impression that blind acceptance of Aristotle's authority has replaced in Averroes' eyes the genuine Aristotelian method of inquiry. Not so. Aristotle's method of instruction consists for him in organizing the *subalternatio* from the universal principles to the particular implementations in the fields of physics and metaphysics, in exactly the same way as he did for the field of logic. In the three main areas of human knowledge—logic, physics, metaphysics—Aristotle had the intellectual power to detect the universal principles and to drive home the research in all the subsets, until each and every discipline that depends upon the relevant overarching science has been explored in detail. This power to reach from universal rules the individual entities is reminiscent of the Neoplatonic model which grants the primary and most effective causality to the most remote cause. The background of Averroes' claim that the mind of Aristotle functioned as a divine rather than a human mind is this model, omnipresent in Arabic philosophy since its beginnings in the circle of al-Kindī.[26] At variance with our minds, that—as remarked, once again, by Aristotle—start from that which is more evident to us but less intelligible per se, Aristotle's divine mind grasped once and forever the principles of science in the various fields. Not only: he was also able to structure the system of knowledge following the ramifications of the departmental sciences, step by step. In doing so, Aristotle systematically followed the demonstrative procedure that is implied in the very notion of science, and the duty of the commentator is to call attention to this. An example taken from Averroes' Long Commentary on the *De Anima* will substantiate this claim.

At the beginning of the *De Anima*, Aristotle assesses the place of the inquiry about soul among the main topics of science:

> Cognition is in our eyes a thing of beauty and worth, and this is true of one cognition more than another, either because it is exact (κατ' ἀκρίβειαν) or because it relates to a more

[25]On the rise and increasing importance of the demonstrative method as the distinctive feature of philosophy and in particular of the works of the "First Teacher" from al-Fārābī onwards see Endress (1990, 1997, 2002); a survey of the *Posterior Analytics* in Arabic philosophy is offered by Marmura (1990).

[26]Once again, lack of space forbids to go further into details: the primacy of the most remote cause alluded to above is stated in the first proposition of the *Liber de causis*, a syllabus of 31 propositions allegedly by Aristotle, but in reality extracted from Proclus' *Elements of theology*, which was translated into Arabic in the so-called "circle of al-Kindī" (Endress 1973). Basic information in D'Ancona and Taylor (2003).

important and remarkable object. On both these grounds we may with good reason claim a high place for the enquiry concerning the soul (*De An.* I 1, 402 a 1–4, trans. Hicks, 3).

The Arabic translation that has come down to us gives the following for this passage:

He said: knowledge (*maʿrifa*) is attained by means of things that possess dignity and nobility, and some prevail over others either by the minute examination of the theory (*istiqṣāʾ al-naẓar*) and the subtlety of the doctrine (*laṭāfa al-maḏhab*), or because their glory and the wonder they induce prevail over others; now, it is our duty to give preliminary information about the soul for both reasons (p. 3.2–4 Badawī, my trans.).

Averroes has at his disposal a different translation, which we can only reconstruct from the Latin, because the Long Commentary on the *De Anima* is lost in Arabic, except for fragments. Here is the text he has before him.[27]

Since knowing something about things differing from one another is considered honorable either in virtue of subtlety (*aut in subtilitate*) or because those are known through things more splendid and more noble, it is right for these two reasons to give the discourse on the soul a position of priority (p. 4 Crawford, trans. Taylor slightly modified).

The two renderings are quite different from one another and their difference is a question we are not compelled to discuss here, but it is a fair guess that the Latin *subtilitas* stands for the Arabic *laṭāfa*, which is the word that features in the extant Arabic translation. Look now at Averroes' commentary on this lemma:

By *subtlety* he means the confirmation associated with demonstration [*certitudo demonstrationis*, **yaqīn al-burhān*].[28] By what he said "or because they are known through things more noble" he means the nobility of the subject. For the arts differ from one another in just one of these two ways, either in the confirmation associated with demonstration or in nobility of the subject, or in both ways. For example, Geometry surpasses Astronomy in the confirmation associated with demonstration, but Astronomy surpasses it in the nobility of the subject. He also said: "It is necessary for these two reasons, etc." That is, because these two features are found in the science of the soul, it is necessary that an account of it take precedence over other sciences. This is obvious to those who give it consideration since the subject of this science is nobler than [those of] others and its demonstration is likewise more certain. He began to give [his] account in this way to lead people to a love of [this] science. His account is in the form of a categorical syllogism. It is as if he says: Because we hold that knowledge concerns things which are honorable and desirable, and that such things surpass one another on account of the confirmation associated with demonstration, on account of the nobility of the subject, or on account of both, as we find in the science of the soul, because in these two features it surpasses other sciences, except for metaphysics, we must hold that the science of the soul comes before the other sciences, and for this reason we placed it in a position of priority among all subjects of inquiry (pp. 3–4 Crawford, trans. Taylor, pp. 1–2, slightly modified).

The Arabic *subtlety*, which can mean 'refinement of the argument', but only as a technical use of a word that literally means 'thinness', 'charm', is explained by

[27] For the sake of space I will omit here to discuss the question of the alternate translation that Averroes mentions here and there in his commentary.

[28] On the topic of *yaqīn al-burhān* in relationship with the theory of demonstration in the Arabic reception of the *Posterior Analytics*, especially by al-Fārābī (d. 950), see Endress (2002) and Black (2006).

Averroes through the precise philosophical notion of reaching the certainty about a given subject-matter via categorical syllogisms, thus matching Aristotle's *ἀκρίβεια*, that Averroes does not have in front of him. Of course Aristotle did not say he would have made use of categorical syllogisms in describing the living beings: I do not want to suggest that Averroes' understanding of Aristotle's *ἀκρίβεια* is accurate. My point is rather that he transforms the lemma he has at his disposal into a doctrine that depends upon his idea of Aristotle's scientific method. One may also mention the fact that while al-Kindī displays little knowledge of or interest in the *Posterior Analytics* (*Kitāb al-burhān*) and their treatment of demonstration, Averroes knows this work very well; but to account for this claim implies a survey of the stages of the Graeco-Arabic translations that we shall skip here, for the sake of space. The main point for the present purpose is that for Averroes the order of the sciences and the demonstrative method are intimately connected to one another, as shown by the passage from the commentary on the *De Anima*: but examples can multiply.

Also in the Latin world the translation by James of Venice (mid-twelfth century) of the *Posterior Analytics* ensured the circulation of the criteria for a body of knowledge to be a science,[29] and it is with this mindset that Albert the Great says *philosophi enim est id quod dicit dicere cum ratione*.[30] Albert did not limit himself to pay lip service to the duty of the philosopher to provide demonstrative knowledge; rather, he embarked on extensive commentaries on Aristotle's works—a practice that was continued by another Dominican theologian, his pupil Thomas Aquinas, and which is less obvious than it may seem at first glance. That *theologians* commented extensively upon Aristotle's logical, physical, and metaphysical works, reserving themselves the right to part company with Aristotle, but also taking for granted that truth is made accessible by inquiry *μετὰ ἀποδείξεως*—*cum ratione*—is worth considering, and anything but trite.[31]

References

Amerini F, Galluzzo G (eds) (2014) A companion to the Latin Medieval commentaries on Aristotle's metaphysics. Brill, Leiden-Boston

[29]On James of Venice's translation of the *Posterior Analytics* cf. Ebbesen (1977), Bloch (2008); on its reception, cf. Corbini (2006).

[30]Albert the Great, *De Quindecim problematibus*, VI. Note however that Albert, in his commentary on the *Physics*, criticizes Averroes' 'divinization' of Aristotle's mind: if one believes that Aristotle was a god, one ought also believe that he never erred; if, on the contrary, one acknowledges he was but a man, one must also think that sometimes he was wrong (VIII, tract. 1, 14), a clear response to Averroes.

[31]Lack of space forbids even to try to make good for the loss of a discussion of Albert the Great's and Thomas Aquinas' position, as remarked with good reason by one of the referees. Recent scholarship, taking into account previous contributions on the broad topic of the scholastic attitude towards Aristotle's criteria for scientific knowledge, include Resnick (2013), Amerini and Galluzzo (2014), and Bydén and Thomsen-Thörnqvist (2017).

Barnes J (1969) Aristotle's theory of demonstration. Phronesis 14:123–152. Revised version In: Barnes J, Schofield JM, Sorabji R (eds) Articles on Aristotle, vol. 1. Science. London: Duckworth, pp 65–87

Barnes J (1981) Proof and the syllogism. In: Berti E (ed) Aristotle on science: the posterior analytics. Proceedings of the 8th symposium Aristotelicum. Padua, Antenore, pp 17–59

Barnes J (1993) Aristotle's philosophy of the sciences. Oxford Studies in Ancient Philosophy 11:225–241

Barnes J (1996) Roman Aristotle. In: Barnes J, Griffin M (eds) Philosophia Togata. II. Plato and Aristotle at Rome. Clarendon Press, Oxford, pp 1–69

Bénatouïl Th (2016) Aristotle and the Stoa. In Falcon A (ed) Brill's companion to the reception of Aristotle in antiquity. Leiden–Boston, pp 56–75

Berti E (1991) Les méthodes d'argumentation et de démonstration dans la Physique (apories, phénomènes, principes). In: De Gandt A, Souffrin P (eds) La Physique d'Aristote et les conditions d'une science de la nature. Vrin, Paris, pp 53–72

Black DL (2006) Knowledge (*'ilm*) and Certitude (*yaqīn*) in al-Fārābī's Epistemology. Arabic Sciences and Philosophy 16:11–45

Bloch D (2008) James of venice and the posterior analytics. Cahiers de l'Institut du Moyen Age grec et latin 78:35–50

Bolton R (1987) Definition and scientific method in Aristotle's posterior analytics and generation of animals. In: Gotthelf A, Lennox JG (eds) Philosophical issues in Aristotle's biology. Cambridge, pp 120–166

Bolton R (1991) Aristotle's method in natural science: physics I. In: Judson L (ed) Aristotle's physics. A collection of essays. Oxford, pp 2–29

Burnyeat M (1981) Aristotle on understanding knowledge. In: Berti E (ed) Aristotle on science: the posterior analytics. Proceedings of the 8th Symposium Aristotelicum. Padua. Antenore, pp 97–139

Burnyeat M (2004) Introduction: Aristotle on the foundation of sublunary physics. In: de Haas FAJ mansfeld j (ed) Aristotle's on Generation and Corruption I. Symposium Aristotelicum, Oxford, pp 7–24

Bydén B, Radovic F (eds) (2018) The Parva naturalia in Greek. Springer, Arabic and Latin Aristotelianism. Supplementing the Science of the Soul

Bydén B, Thomsen-Thörnqvist Ch (eds) (2017) The Aristotelian tradition. Aristotle's works on logic and metaphysics and their reception in the middle ages. Brepols, Turnhout

Corbini A (2006) La teoria della scienza nel XIII secolo. I commenti agli Analitici secondi, Sismel—Edizioni del Galluzzo, Firenze

D'Ancona C, Taylor RC (2003) Le Liber de causis. In: Goulet R (ed) Dictionnaire des philosophes antiques, I. CNRS Editions, Paris, pp 549–647

De Haas FAJ (2002) Modifications of the method of inquiry in Aristotle's physics I.1. An essay on the dynamics of the ancient commentary tradition. In: Leijenhorst CH, Leijenhorst C, Lüthy CH, Thijssen JMMH (eds) The dynamics of Aristotelian natural philosophy from Antiquity to the seventeenth century. Brill, Leiden, pp 31–56

De Haas FAJ, Leunissen M, Martin M (eds) (2010) Interpreting Aristotle's posterior analytics in late antiquity and beyond. Brill, Leiden-Boston

Devereux P, Pellegrin P (eds) (1990) Biologie, logique et métaphysique chez Aristote. CNRS Editions, Paris

Donini P (1994) Testi e commenti, manuali e insegnamento: la forma sistematica e i metodi della filosofia in età postellenistica. In: Haase W, Temporini H (eds) Aufstieg und Niedergang der römischen Welt II.36.7. De Gruyter, Berlin–New York, pp 5017–5100

Düring I (1957) Aristotle in the ancient biographical tradition, Almqvist & Wiksell, Göteborg & Stockholm

Ebbesen S (1977) Iacobus veneticus on the posterior analytics and some early 13th Century Oxford masters on the Elenchi. Cahiers de l'Isntitut du Moyen Age grec et latin 21:1–9

Ebrey D (ed) (2015) Theory and practice in Aristotle's natural science. Cambridge U.P, Cambridge

Endress G (1973) Proclus Arabus. Zwanzig Abschnitte aus der Institutio Theologica in arabischer Übersetzung. Imprimerie Catholique, Wiesbaden-Beirut

Endress G (1990) The defence of reason: the plea for philosophy in the religious community. Zeitschrift für Geschichte der Arabisch- Islamischen Wissenschaften 6:1–49

Endress G (1997) L'Aristote arabe. Réception, autorité et transformation du 'Premier Maître', Medioevo 23:1–42

Endress G (2002) The language of demonstration: translating science and the formation of terminology in Arabic philosophy and science. Early Science and Medicine 7:231–254

Falcon A (2005) Aristotle and the science of nature: unity without uniformity. Cambridge U.P, Cambridge

Falcon A (2016) Introduction. In Falcon A (ed) Brill's Companion to the Reception of Aristotle in Antiquity. Leiden–Boston, pp 1–9

Falcon A (2017) Aristotelismo. Einaudi, Torino

Ferejohn MT (1990) The origins of Aristotelian science. Yale U.P, New Haven

Gotthelf A (2012) Teleology, first principles, and scientific method in Aristotle's biology. Oxford U.P, Oxford

Gotthelf A, Lennox JG (eds) (1987) Philosophical issues in Aristotle's biology. Cambridge U.P, Cambridge

Gottschalk HB (1987) Aristotelian philosophy in the Roman world. In: Haase W, Temporini H (eds) Aufstieg und Niedergang der römischen Welt II.36.2. De Gruyter, Berlin–New York, pp 1079–1174

Grote G (1880) Aristotle. Edited by Robertson GC, Bain A, 2nd ed. with additions. Murray, London

Hadot I (1987a) La division néoplatonicienne des écrits d'Aristote. In: Wiesner J (ed) Aristoteles. Werk und Wirkung. Paul Moraux gewidmet. II: Kommentierung, Überlieferung, Nachleben. De Gruyter, Berlin, pp 249–285

Hadot I (1987b) Les introductions aux commentaires exégétiques chez les auteurs néoplatoniciens et les auteurs chrétiens. In: Tardieu M (ed) Les règles de l'interpretation, Cerf, Paris, pp 99–122

Hadot I (1991) The role of the commentaries on Aristotle in the teaching of philosophy according to the prefaces of the Neoplatonic commentaries on the categories. Oxford Studies in Ancient Philosophy, Supplementary Volume. In: Blumenthal HJ, Robinson H. Clarendon Press, Oxford, pp 175–189

Hadot P (1979) Les divisions des parties de la philosophie dans l'Antiquité. Museum Helveticum 36:202–223

Hadot P (1987) Théologie, exégèse, révélation, Écriture dans la philosophie grecque. In: Tardieu M (ed) Les règles de l'interprétation. Cerf, Paris, pp 13–34

Hahm DH (1991) Aristotle and the stoics: a methodological crux. Geschichte der Philosophie 73:771–791

Hamlyn DW (1990) Aristotle on dialectic. Philosophy 65:465–476

Harvey S (1985) The Hebrew translation of Averroes' Prooemium to his long commentary on Aristotle's physics. Proceedings of the American Academy for Jewish Research 52:55–84

Hatzimichali M (2016) Andronicus of Rhodes and the construction of the Aristotelian corpus. In: Falcon A (ed) Brill's companion to the reception of Aristotle in antiquity. Leiden–Boston, pp 81–100

Hein Ch (1985) Definition und Einteilung der Philosophie. Von der spätantiken Einleitungsliteratur zur arabischen Enzyklopädie. P. Lang, Frankfurt, Bern-New York

Hugonnard-Roche H (2007) Le corpus philosophique syriaque aux VIe-VIIe siècles. In: D'Ancona C (ed) The libraries of the Neoplatonists. Proceedings of the Meeting of the European Science Foundation Network "Late Antiquity and Arabic Thought. Patterns in the Constitution of European Culture", Brill, Leiden–Boston, pp 279–291

Lennox JG (2001) Aristotle's philosophy of biology: studies in the origins of life science. Cambridge University Press, Cambridge

Lloyd GER (1990) The theories and practices of demonstration in Aristotle. The Proceedings of the Boston Area Colloquium in Ancient Philosophy 6, 371–401. Reprinted with additional notes

and a Postscriptum in Lloyd GER (1996) Aristotelian explorations. Cambridge University Press, Cambridge, pp 7–36
Lynch JP (1972) Aristotle's school. A study of a Greek educational institution. University of California Press, Berkeley
Marmura ME (1990) The fortuna of the posterior analytics in the Arabic middle ages. In: Astzalos M, Murdoch J, Niinliuoto I (eds) Knowledge and the sciences in medieval philosophy. Proceedings of the Eighth International Congress of Medieval Philosophy. S.I.E.P.M., Helsinki, pp 85–103
McKirahan RD (1978) Aristotle's subordinate sciences. The British Journal for the History of Science 11:197–220
McKirahan RD (1992) Principles and proofs. Aristotle's theory of demonstrative science. Princeton U.P., Princeton
Moraux P (1973) Der Aristotelismus bei den Griechen von Andronikos bis Alexander von Aphrodisias. Band 1. Die Renaissance des Aristotelismus im 1. Jh. v. Chr. De Gruyter. Berlin
Owen GEL (1961) *Τιθέναι τὰ φαινόμενα*. In: Mansion S (ed) Aristote et les problèmes de méthode. Symposium Aristotelicum. Louvain, pp 83–103. Reprinted In: Moravcsik JM (ed) (1967) Aristotle. A collection of critical essays. Doubleday, New York, pp 167–190. In: Barnes J, Schofield JM, Sorabji R (eds) (1975) Articles on aristotle, vol 1, Science, pp 113–126
Primavesi O (2007) Ein Blick in den Stollen von Skepsis: Vier Kapitel zur frühen Überlieferung des Corpus Aristotelicum. Philologus 151:51–77
Resnick IM (ed) (2013) A companion to Albert the great. Theology, Philosophy, and the Sciences. Brill, Leiden–Boston
Sandbach FH (1985) Aristotle and the Stoa. Cambridge Philological Society, Cambridge
Sassi MM (2014) Percezione e conoscenza nei Parva naturalia. Studia graeco-arabica 4:265–274
Scharle M (2015) "And these things follow": Teleology, necessity, and explanation in aristotle's meteorologica. In: Ebrey D (ed) Theory and practice in aristotle's natural science. Cambridge U.P., Cambridge, pp 79–99
Sedley D (1996) Plato's auctoritas and the rebirth of the commentary tradition. In: Barnes J, Griffin M (eds) Philosophia Togata. II. Plato and Aristotle at Rome. Clarendon Press, Oxford, pp 110–129
Wians W (1989) Aristotle, demonstration, and teaching. Ancient Philosophy 9:245–253
Wieland W (1960–1961) Das Problem des Prinzipienforschung und die aristotelische Physik. Kant-Studien 52:206–219. English version: Aristotle's Physics and the Problem of Inquiry into Principles. In: Barnes J, Schofield JM, Sorabji R (eds) Articles on Aristotle, vol. 1. Science. London, Duckworth, pp 127–140
Wilson M (2013) Structure and method in Aristotle's meteorologica: a more disorderly nature. Cambridge U.P, Cambridge

Primary Sources

Albertus Magnus, Albertus Magnus, Physica, V-VIII edidit P. Hossfeld, in Alberti Magni Opera Omnia IV/1, Aschendorff, Münster 1993
Albertus Magnus, De XV Problematibus edidit A. Hufnagel, in Alberti Magni Opera Omnia XVII/1, Aschendorff, Münster 1975
Alexandri in Aristotelis Meteorologicorum libros commentaria (...) edidit M. Hayduck, Reimer, Berlin 1899 (CAG III, 2)
Aristotle De Anima recognovit (...) W.D. Ross, Clarendon Press, Oxford 1956
Arabic Translation: ʿA. Badawī (ed.), Arisṭūṭālīs fī l-nafs. al-Ārāʾ al-ṭabīʿiyya al-mansūb ilā Fulūṭarḫus. al-Ḥāss wa-l-maḥsūs li-Ibn Rušd. al-Nabāt al-mansūb ilā Arisṭūṭālīs, Maktabat al-nahḍa al-miṣriyya, Cairo 1954. Reprint Kuwayt/Beirut 1980, esp. pp. 3–88 (= Kitāb Arisṭāṭālīs wa-naṣṣ kalāmihī fī l-nafs. Tarǧamat Isḥāq ibn Ḥunayn)

Aristotle Meteorologica with an English Translation by H.D.P. Lee, Harvard U.P.–Heinemann, Cambridge (MA) London 1952

Aristotelis Opera cum Averrois commentariis, Venetiis apud Junctas, 1562–1574 (reprin. Minerva, Frankfurt a.M. 1962), vol. IV, Aristotelis de Physico auditu libri octo cum Averrois Cordubensis (...) commentariis

Averrois Cordubensis, Commentarium magnum in Aristotelis De Anima libros recensuit F.S. Crawford, Cambridge [MA] 1953 (Corpus Commentariorum Auerrois in Aristotelem ... Versionum Latinarum VI 1)

Averroes (Ibn Rushd) of Cordoba, Long Commentary on the De Anima of Aristotle. Translated and with Introduction and Notes by R.C. Taylor with Th.-A. Druart, subeditor, Yale U.P., New Haven–London 2009

Ioannis Philoponi In Aristotelis Meteorologicorum librum primum commentarium (...) edidit M. Hayduck, Reimer, Berlin 1901 (CAG XIV, 1)

Kupreeva I (2011) Philoponus: on Aristotle meteorology 1. 1–3. Translated by I. K., Bloomsbury, 507 London–New York

Rasā'il al-Kindī al-falsafiyya, Dār al-fikr al- 'arabī, vol. I, Cairo 1950

The Philosophical Works of al-Kindī, P. Adamson and P. Pormann (trans.), Oxford U.P., Oxford–New York 2012

Philosophy and Scripture. Medieval Jewish Opinions on Creation and the Laws of Nature

Elisa Coda

Abstract This paper deals with one of the most controversial issues in Medieval thought: the question of whether or not creation out of the will of God entails occasionalism. An overview is provided of the approach to this topic by Isaac Abravanel (1437–1508), a polymath and a key figure in Jewish thought. Focusing on Abravanel's assessment of creation out of nothing and annihilation of the cosmos in his theological treatises, and parallel assessment of an Aristotelian cosmos in his *Forms of the Elements*, this paper provides an example of Jewish thought about Scripture and Aristotelian philosophy between the Middle Ages and the Renaissance.

Keywords Isaac abravanel · Eternal creation hypothesis · Occasionalism

1 The Problem

The intellectual terrain over which Isaac Abravanel (1437–1508)[1] trod is complex. His life spans between the Middle Ages and the Early Renaissance.[2] Abravanel's cosmology stands at the very end of a millenary cosmological controversy, just before the ancient-medieval conception of the cosmos was upset by the scientific revolution. Yet, it might be useful to propose a survey of the main questions about creation rooted in Greek cosmological speculations, and still having their defenders

[1]On Abravanel's life and works see e.g. Netanyahu (1953), Feldman (2003). See also Goetschel (1997), Lawee (2001).

[2]Scholars disagree about the appropriate historiographical framework of Abravanel's intellectual outcome. An assessment of this debate is carried out in Kellner (1982a, b), Cohen Skalli (2007, 2009), Lawee (2009), Coda (2018), 1–43.

My most sincere thanks go to Prof. Adriano Fabris and all the participants of the research project "Ethics, Science, and Democracy" (University of Pisa) that has granted me the opportunity to carry on this research.

E. Coda (✉)
University of Pisa, Department of Civilizations and Forms of Knowledge, via dei Mille 19, 56125 Pisa, Italy
e-mail: elisa.coda@cfs.unipi.it

A. Fabris (ed.), *Trust*, Studies in Applied Philosophy, Epistemology and Rational Ethics 54, https://doi.org/10.1007/978-3-030-44018-3_3

among the *falāsifa* and Jewish thinkers during the Middle Ages. This will provide us with as a rough map of Abravanel's complex intellectual heritage made of centuries of cosmological controversies, philosophical speculations and exegesis of the Bible.

From Late Antiquity to the Middle Ages, the creation of the world has been one of the most important topics discussed in theological, philosophical and scientific writings by pagan, Christian, Muslims, and Jewish authors. What is the precise dogmatic status of the doctrine of creation? Does creation feature as a cosmological hypothesis that can be philosophically proven, or one must have recourse to revelation to assess that the world was created by God? And does creation imply destruction of the world? In medieval cosmological thought these issues are linked together, because questions on the origin of the world entail questions about God. Put otherwise, cosmology reflects theology and is linked to it (Wolfson 1961, 1971, 1976). For philosophers, the question of the origin of the world exceeds the boundaries of the notion of 'creation', as one can easily realize if one thinks of Aristotle's objections to Plato's *Timaeus*. Theologians, or philosophers operating in a context of creationism theologically based—be it Christian, Muslim, or Jewish—often saw the question of creation as fundamental, entailing as it does conclusions about God's nature, goodness, immutability, and power. Another issue, and one that will prove decisive for Abravanel, is that of the end of the world (which in turn entails conclusions about the world's ontological status). If the universe has a beginning in God, God himself could constitute a sufficient condition for its eventual demise. Étienne Gilson saw in *Genesis* I 1 a hint to the metaphysical novelty of this kind of radical contingecy of beings subject to becoming. According to Gilson, Thomas Aquinas' *philosophical* idea of identity of essence and existence in God and nowhere save in God entails the following conclusions about the ontological status of the world:

> Nothing could be more familiar than the first verse of the Bible: "In the beginning God created the Heavens and the Earth" (Gen. I 1). Here once more we have no trace of philosophy. God asserts His creative action, just as He asserts the definition of His being, without any kind of metaphysical justification. And yet between these two unproved assertions, how profound, how inevitable is the metaphysical accord! If God is Being, if He alone is Being, the all that is not God must of necessity hold its existence from God (Gilson 1940, 68).

Gilson was perfectly aware of the importance of Arab-Muslim philosophy in establishing the topic of the *contingentia mundi*, whose creator was for him Avicenna.[3] But in Avicenna the contingency of the world went hand in hand with the necessity of its derivation from the Necessary Being, whose 'action' to give rise to contingent beings was nothing more than the consequence of its nature, as the radiance is a consequence of the nature of the sun. As is well known, this position was harshly criticised by al-Ghazālī, and the latter's condemnation of Avicenna (and for that matter of the whole of *falsafa*, the Graeco-Arabic philosophy) was, in its turn, criticised by Averroes. No matter how rapidly we remind ourselves of this background, it is useful to replace ourselves in the context of the discussion about

---

[3]Avicenna's debt to late ancient philosophy, especially John Philoponus, was not especially remarked by Gilson.

Scripture on the one hand, and the 'Aristotelian' cosmos on the other, that characterizes Jewish thought between the Middle Ages and the Renaissance. Indeed, the conceptual schemes roughly summarized above are still spread within intellectuals of Spanish and Italian Jewry in Abravanel's lifetime.

2 Isaac Abravanel: Is Creation a Dogmatic Tenet?

The doctrine of creation is pervasive in Abravanel's multifaced litterary production. Let me begin by talking about the dogmatic status he attributes to this doctrine. The research of Seymour Feldman, Menachem Kellner, and more recently Eric Lawee and others has given us much information of the nature, ambivalence and development of Abravanel's emphasis on the dogmatic importance of belief in creation as a foundamental principle in Judaism. Indeed, following in Maimonides' footsteps,[4] Abravanel stresses the doctrinal importance of belief in creation, that he repeatedly posits as the foundation of the whole Torah. Thus, for example, he writes:

> Where I do to chose principles to posit for the divine Torah I would only lay down one, the creation of the world. It is the root and the foundation around which the divine Torah, its cornerstones, and its beliefs revolve, and it includes the creation at the beginning, the narratives about the patriarchs, and the miracles and wonders which cannot be believed without belief in creation. The same applies to belief in God's knowledge and providence, and in reward and punishment according to [one's observance of] the commandments, none of which can be perfectly believed in without believing in the volitional creation of the whole world.[5]

As Kellner showed several decades ago, Abravanel's claim that creation is a fundamental principle of Judaism is consistently widespread in his writings but contradicts his belief that "it is not proper to postulate principles for the divine Torah, nor foundations in the matter of beliefs, for we are obliged to believe everything that is written in the Torah. (...) For, since the Torah is true, no belief or narrative in it has any advantage over any other.[6]" Even if Abravanel seems not to be entirely consistent in this regard, he reiterates the position of creation as axiomatic to Judaism in two other works, *New Heavens* (*Shamayyim Chadashim*) and *Deeds of God (Mif 'a lot Elohim)*,[7] where he writes that "the creation of the world is the foundation (*yesod*) of the Torah".[8]

---

[4]See e.g. *Guide of the Perplexed*, Part II, Chaps. 13, 25, and 30.

[5]*Principle of Faith (Rosh Amanah)*, Chap. 22, p. 192 (trans. By M. Kellner).

[6]*Isaac Abravanel. Principle of Faith (Rosh Amanah)*, Ch. 23, pp. 196–197 (trans. By M. Kellner).

[7]See below, pp. 00–00.

[8]*New Heavens* (*Shamayim Chadashim*), 1: fol. 2v; bibliographical references below, n. 13. See also *Deeds of God (Mif 'a lot Elohim)*, 1: 3, fol. 6r col. a line 5.

Creation is not only a dogmatic tenet in Abravanel's account, but also something that can be proved by demonstrative arguments.[9] Distinguishing from a thing that is impossible in itself (i.e. by intrinsic impossibility), and another that is impossible for some agent (i.e. by relative impossibility), Abravanel says:

> The existence of something corporeal from no antecedent matter is not absolutely impossible, for it is not something whose conception is self-contradictory (...) Rather, the generation of something corporeal from nothing is impossible relatively, i.e., with respect to some natural agent or craftsman, since [in the latter cases] their power is finite, and hence cannot act except upon some subject [i.e. matter].[10]

Creation means, in Abravanel's account, a unified system whose parts are all ordered by God's will:

> There is no doubt that we cannot say with respect to the entire system that this [integration of the microsystems] results from the power of their own nature. For, although each one of them acts according to its own nature, [we must ask] who put them together in such a way that their actions agree with each other to bring forth in cooperation that benefit? [This question cannot be answered] unless we posit one thing outside these systems that arranged them in a unified, cooperative act or actions. It is this [cause] that gave them that unity by virtue of which the benefit and good [in the world] derives.[11]

Thus, creation seems to have a double status in Abravanel's thought: on the one side, it is a belief firmly rooted in Scripture; on the other, its truth results from a refined, critical dialogue with the typical arguments of *Kalām*, as it happens so often also in Maimonides.[12] The first passage above, where creation out of nothing is accounted for through the idea that divine power is infinite, echoes the adoption in the *Kalām* of a topic ultimately derived from Philoponus, but whose direct source in Abravanel is in all likelihood Maimonides. It was Philoponus who transformed the power of the Aristotelian Unmoved Mover to impart infinite movement to the celestial spheres into God's creation of being out of His infinite power.[13] However, the form this argument takes in Abravanel is clearly reminiscent of Maimonides' developments.[14] The second passage, where the mutual benefits of the parts of the created cosmos suggest the existence of a causal and provident principle, echoes the argument from design, typical of both *falsafa* and *Kalām*.[15] Far from being surprising, this is exactly what one would expect from a scholar with a solid background in that kind of philosophy and philosophical theology that was widespread in the multicultural context of the Iberian Peninsula at the time of Abravanel's education, and whose dialogue with the Jewish faith was established once and for all by Maimonides.

---

[9] Abravanel objects to the Maimonidean thesis according to which the question of how the world began is not susceptible to demonstration. About Maimonides' account of creation see Seeskin (2005), Ivry (2016), 86–122.

[10] *Deeds of God*, 4: 2, fol. 31r-v (trans. Feldman 2003, 51 modified).

[11] *New Heavens* (*Shamayim Chadashim*), 3: 31r (trans. Feldman 1990, 23).

[12] See Ivry (2016).

[13] Davidson (1987), 86–116; Davidson (1969), cf. also Feldman (1990).

[14] Ivry (2016), 88–89.

[15] Davidson (1987), 117–153 and 213–236, Sorabji (1987).

What is less obvious is that this created world is an Aristotelian one. Of course, the adjective 'Aristotelian' points here less to Aristotle himself than to the Aristotle of the Graeco-Arabic-Jewish tradition, i.e. to Aristotle as translated and adapted since the beginning of Arabic philosophy.[16] It is nevertheless remarkable that Abravanel's description of the physical world pivots on the exact interpretation of Aristotle's doctrine of the nature of matter and interplay of the elements.

3 An Aristotelian Sublunar World

Between 1457 and 1467, when he was still in Portugal, Isaac Abravanel wrote*The Forms of the Elements*,[17] a treatise devoted to discussing the interpretation of Aristotle's *On Generation and Corruption* and rooted in the multisecular tradition of commenting upon this work. The question raised in Abravanel's treatise has directly to do with the nature of the visible world: what exactly are its primary elements, how do they cooperate in producing the things that are grasped by sense-perception, and their behaviour? Are the primary elements of the physical world, i.e. the four of the Aristotelian tradition—fire, water, air, earth—primary qualities, or rather should they be reduced to the couple of heavy and light, or again, do they belong as properties to first matter, which is in itself indeterminate? After a detailed excursus of each of these three positions, articulated into objections that can be raised to each, and have actually been raised by a series of authors[18] including Alexander of Aphrodisias, Themistius, Avicenna, al-Ghāzālī, Ibn Tufayl, Averroes, Gersonides, and Moses Narboni, Abravanel comes to the conclusion that

> There is a quality that is typical of each [element] and is primary, if compared to what is present in another element: if heat is the form of fire, wet is the form of water, cold is the form of air, and dry is the form of earth, and the other qualities that are present in them are not their forms, because each of these elements possesses other features in the way which is appropriate to it, it follows that heat, when it is considered in air, is not as it is when it is present in fire; and wet, when it is considered in the water, is not as it is when it is considered in air, and cold, when it is considered in earth, is not as it is when it is considered in water. Nor heat, when it is considered in air, is as it is when it is considered in water; and the cold of the earth is not of the same species of the heat of fire, or even of the wet as considered in the air, or of the cold of the water. As a matter of fact, even if all these are called "qualities", in their actual relationship they change with respect to their nature. Since there are both qualities per se and secondary qualities, the features that such things possess are not "forms", as I have

[16]See R. Sorabji (ed.), *Aristotle Transformed. The Ancient Commentators and Their Influence*, Cornell U.P., Ithaca N.Y 1990 (rist. Bloomsbury Academic, London 2016); Id (ed.), *Aristotle Re-Interpreted. New Findings on Seven Hundred Years of the Ancient Commentators*, Bloomsbury Academic, London 2016.

[17]This is the first work by Abravanel; the *editio princeps* was published in Sabbioneta (Italy) as an appendix to *The Crown of the Ancients* (*'Ateret Zeqenim*).

[18]Coda (2018), 49–70.

demonstrated. It follows from this that the primary qualities are the forms of the elements, and this is what I had set for myself to explain.[19]

After detailed examination of all the possible interpretations of the basic structure of bodily natures, one description appears to meet the requirements of scientific truth: that which identifies in the primary qualities the forms that constitute the nature of each element. One should be careful not to confuse a perceptible quality, e.g. the heat of hot water with the form, in this case Heat: if considered in fire, Heat is its form, while if considered in hot water, heat is a quality. Roughly speaking, this is how Abravanel understands both the structure of the visible world, and Aristotle's account of it.

All in all, Aristotle's *On Generation and Corruption*, once analyzed as Abravanel does in his treatise, offers a viable description of what the elementary structures of the sublunar world are, and how they function. Nowhere in the *Forms of the Elements* he feels obliged to compare the Aristotelian world with God's will or the Scriptures: the nature of the elements does not depend upon either. The question of the beginning of the universe is not raised and, what is more, Abravanel keeps silent about the very fact that the Scriptures provide a different account of the constitution of the visible world. A simple explanation springs to mind: we know that the *Forms of the Elements* is an early work, written when Abravanel was still in Portugal, and one may think of it as of a sort of scholarly exercise in Aristotelian exegesis, with no precise theological import. In order to check this explanation, let's now turn to the three works that a mature Abravanel composed on the issue of creation, when he moved to Italy.

4 Creation: The Italian Trilogy

Readers of Abravanel's most celebrated works are accustomed to apocalyptic visions and messianic speculations. Such writings however side with the philosophical analysis of the beginning of the universe. During his stay in the Italian town of Monopoli (1495–1503), at that time under Venetian control, and not too long after the completion of his so-called 'messianic trilogy'[20]—*The Fountains of Salvation* (*Ma 'ayenei Yeshu 'at*, 1497), *The Salvation of His Annointed* (*Yeshu 'ot Meshico,* 1497) and *The Announcer of Salvation* (*Mashmi 'a Yeshu 'ah* 1498)—Abravanel began a new trilogy. The focus this time was on creation. First came the *New Heavens* (*Shamayyim Chadashim*, 1498),[21] then one of Abravanel's main philosophical treatises, the *Deeds*

[19]Coda (2018), 107.3–13.

[20]See Netanyahu (1953), 77–78, Feldman (2003), 9–10, 128–149.

[21]*Sefer Shamayim Chadashim* (1828), repr. (1972) —in the present article, references will be given from Coda (2018). The treatise contains four parte: *Shamayim Chadashim*, 1: fols 1r-12v; 2: fols 13r-23r; 3: fols 23v-34r; 4: 34v-51r. The first one should be—at least partially—regarded as a commentary on chapter 19 of Part II of Maimonides' *Guide of the Perplexed* where Maimonides

of God (*Mif'alot Elohim*, 1501), and eventually a *Commentary on Genesis* 1:1 (*Commentary to the Torah*, 1506–1507). Notwithstanding its importance, I shall ignore here the escathological side of Abravanel's thinking, narrowing the focus to the philosophical arguments for the beginning of the universe, and to some cosmological aspects entailed by them.[22]

The basic question is: if God created the world out of no matter at the first istant of time, and if the universe has a beginning, does creation imply that the universe is also subject to destruction, and will have a last instant? Answering in the affirmative, as it seems inevitable, entails that if so, it will be God who will destroy it, deciding when the last instant will occur. Abravanel's position, combining philosophical and theological motifs, is that the world is obviously destructible; time had a first instant, and will have a last one.[23] He strongly rejects the doctrine of eternal creation.[24] The only necessary existing substance is God; He created everything else as essentially contingent and finite. This world will be destroyed, and another one will be created, having the same nature as a finite being. As a consequence, God can create, and indeed creates, a plurality of finitely enduring worlds.[25] Yet, the way in which, according to Abravanel, God operates the destruction of this world and the creation a new one does not entail any change in God Himself. Since God is a voluntary agent whose will is identical with his knowledge, He created successively a plurality of universes out of love. The intrinsic destructibility of the world does not imply any change in God's will. On the contrary, destructibility depends upon the opposite items that are present in the material world; even though heavenly bodies are incorruptible, this does not mean that they will not perish in the end. It is true that they

> do not undergo corruption according to the natural standard, which is from one opposite to the other opposite. Rather, just as we believe that the heavens came into being neither from opposites nor from the elements, but from no-thing and after complete privation, so too will they perish, not into the opposite and not into the elements, but into complete privation (*Deeds of God*, trans. by Ogren 2015, 148).

One may surmise that this is a confuse statement. In reality, the basic idea is that the two causative acts by which God creates and annihilates the world are radically different from any other causal relationship in the created world. The 'change' imparted by God cannot be compared with any other change occurring in the created world. Hence, after that creation has been made, and until the created world will be annihilated by God, this created world is entitled to function according to its

---

explains how, independently of the authority or Scriptures, he has been led to adopt the belief in the creation *ex nihilo*.

[22]See e.g. Netanyahu (1953), 195–254, Feldman (2003), Chaps. 10–12; Strauss (2013).

[23]According to Feldman (1986, 1990), Abravanel makes use of *kalāmic* materials taken from Saadia Gaon's (882–942 CE) arguments for creation *ex nihilo* as in the *Book of Beliefs and Opinions*. He also refers to John Philoponus' refutation of the Aristotelian thesis of the eternity of the world in his *De aeternitate mundi*.

[24]A similarity of Abravanel's approach on the matter to those of Alemanno has been suggested by Ogren (2015).

[25]See below, pp. 00–00.

own rules, with heavenly bodies rotating regularly forever—i.e., as long as there is time—and the sublunar elements moving from one opposite to the other opposite, combining together, coming-to-be and passing over: all the kinds of change that we can see in the celestial and sublunar worlds have nothing to do and are not comparable to creation and annihilation. At variance with creation and annihilation, the changes in the created world can be described according to the Aristotelian science. As for the sublunar world, in the *Forms of the Elements* Abravanel change is described as follows:

> As Aristotle says, the things that change due to a change in form can change, due to the change of form, also from the point of view of the place: for example, in the case of heat or cold, what comes close to the upper extremity is hotter, while it is colder that which approaches the center; or, again, the first act of the life of a living being is elementary heat, as explained in the book *On Generation and Corruption*, and in other places. All this shows that heat and cold are the forms of the elements.[26]

The trouble is that nothing in the "Italian Trilogy" suggests that Abravanel revised this description, inspired by Aristotle and working a with the hypothesis of regular laws of nature. Even in these later works, Abravanel does not seem to think that accepting—with Aristotle—the existence of laws of nature that regulate the normal operation of sublunary bodies is in conflict with the omnipotence of God. Indeed, the absolute power of God manifests itself in creating and destroying only. In the *Deeds of God* he writes:

> The nature of everything is circular[27] for everything is destined to return to the place from which it came. If it has come from the elements, it will return to them, and if it has come from prime matter, it will come back to that; and if it has come from corruption, to that it will come back in any case. And if something has emerged and has emanated from Him, Blessed be he, it will be gathered in Him, according to the saying of Solomon: "and dust returns to the earth as it was, and the spirit returns to God who gave it (Qo. 12:7)." For behold! The creation of the world and its regeneration are the Torah.[28]

In the "Italian Trilogy" neither conflict nor harmony are claimed between the Scriptures and the Aristotelian account of how nature functions, be it in the heavenly or in the sublunar worlds. Abravanel seems to operate with the assumption that divine omnipotence has to do with the beginning and end of the universe, whereas during its functioning the created world works according to Aristotelian rules.

5 A Created, yet 'Aristotelian' World

Let me turn finally to a theological implication of Abravanel's philosophical discussions about creation. I have suggested in the previous paragraphs that Abravanel

---

[26] *Forms of the Elements*, p. 105.16–21 ed. Coda. For an in-depth analysis of Abravanel's discussion about corporeal forms see Elior (2012).

[27] For the "cosmic circles" in Abravanel see Ogren (2006–2007).

[28] Quoted in Ogren (2006–2007), 152–153.

posits creation *ex nihilo*—implying the absolute beginning of the world from no pre-existent eternal matter—as the basic dogma in Judaism. But positing a beginning instant carries with it assumptions about what preceded this instant, and expecially its representation in the Scriptures. This problem became the focus of philosophical speculation among Jewish thinkers wishing to ground Scripture in philosophic or scientific legitimacy. A number of pressing questions emerged from this issue. Which is the ontological status of the created world? Could there be any laws of nature?

Indeed, Abravanel makes the existence of thelaws of nature compatible with any suspensions of natural necessity caused directly by God's will. This issue is discussed at length in the "Italian Trilogy",[29] where he discusses both the extraordinary character of divine interventions altering the natural necessity in the course of events, and the fact that such unnatural features occur for a purpose.

> Indeed, a miraculous event (*ha-davar ha-nisi*) is that which is generated without its natural causes (...) it is the existence of something without its natural causes and dispositions, performed with intention by someone with will.[30]

The same argument that is at work in the accounts of miracles can be adopted also to account for celestial anomalies:

> For if the master were asked about the celestial anomalies and he were to determine that they cannot be explained by monic necessity but only by means of will and intention, that is from a divine will that intended to make them in this way, the unification method has been demonstrated. For God has created them according to that agreement, relation, and unity which they exhibit.[31]

To rephrase this discussion in terms more familiar to the Latin tradition, Abravanel seems put it close to the position that establishes the mutual consistence of laws of nature *de potentia Dei ordinata*, and the logical possibility, *de potentia Dei absoluta*, to alter such laws—a logical possibility that finds confirmation *a parte post* in the narrative of miracles made in the Scriptures.

To what extent al-Ghazālī's well-known denial of extra mental causal connections in the *Destruction of the Philosophers*[32] entails occasionalism is a long-debated question,[33] and I will not even attempt to enter the debate. However, it is useful to have ready at hand the well-known passage where extra mental causal connections are denied:

> The connection between what is habitually believed to be a cause and what is habitually believed to be an effect is not necessary, according to us. But [with] any two things, where 'this' is not 'that' and 'that' is not 'this' and where neither the affirmation of the one entails

[29]Thus, Abravanel ends the *Deeds of God* with a detailed account of miracles. See the in-depth analysis by Feldmann (2003), 67–89, with an overview of the main critiques that Abravanel made to Maimonides' and Gersonides' treatments of miracles.

[30]*Deeds of God (Mif 'a lot Elohim)*, 10:10, fol. 97v, trans. in Feldmann (2003), 70.

[31]*New Heavens (Shamayim Chadashim)*, 3, fol. 32r-v, trans. in Feldmann (1990), 24.

[32]Marmura (2000), 166–225.

[33]Marmura (2005), 145–152. See also Fakhry (1958, repr. 2007), 56–82, Goodman (1978), Perler-Rudolph (2000), 75–77.

> the affirmation of the other nor the negation of the one entails negation of the other, it is not a necessity of the existence of the one that the other should exist, and it is not a necessity of the nonexistence of the one that the other should not exist—for example, the quenching of thirst and drinking, satiety and eating, burning and contact with fire, light and the appearance of the sun, death and decapitation, healing and the drinking of medicine, the purging of the bowels and the using of a purgative, and so on to (include) all [that is] observable among connected things in medicine, astronomy, arts, and crafts. Their connection is due to the prior decree of God, who creates them side by side, not to its being necessary in itself, incapable of separation. On the contrary, it is within [divine] power to create satiety without eating, to create death without decapitation, to continue life after decapitation, and so on to all connected things. The philosophers denied the possibility of [this] and claimed it to be impossible (al-Ghazālī. *The Incoherence of the Philosophers*. Seventeenth Discussion—trans. Marmura 2000, 166).

Mention of al-Ghazālī, is made necessary also because no less an acute interpreter of Abravanel's theological position as Seymour Feldman suggests that he is inspired by al-Ghazālī in his treatment of miracles as exceptions to laws of nature, which implies that there are indeed such laws, but God has the power to suspend them.[34] Though agreeing with the fact that this is the way in which Abravanel makes divine creation-and-annihilation consistent with his 'Aristotelian' cosmos, it seems to me that Abravanel's position differs from al-Ghazālī's account as for the status of extramental causal connections is concerned. In Abravanel's eyes, the four elements of the sublunar world really interact with each other. Their interactions really cause phenomena. Or at least nothing suggests that when dealing with the topic of creation-and-annihilation of the world, or after he had dealt with it, Abravanel renounced describing the visible world in this way, along with 'Aristotle' and the Aristotelian tradition. This stance seems to be inspired at least in part by Maimonides. As pointed out by Ivry,

> Maimonides believes that the philosophers' proofs of God's existence and nature remain valid whether the physics of the universe are coeternal with God or are subsequent to His creation. Maimonides' confidence in the nature and de facto permanence of the natural world order is unassailable. His conceptualization of God follows naturally from his physics and sense of logical imperatives and is unimaginable apart from the world (however separate and ultimately unimaginable He purportedly is).[35]

While the first part of Ivry's overview of Maimonides' philosophical theology applies in full also to Abravanel, the second part could no longer function as a description of how Abravanel sees the relationship between God and the world, at least in the "Italian Trilogy". An Aristotelian cosmos, in both the celestial movements and sublunar elements, had been created and will surely be annihilated. The tension between the laws of nature that are surely at work as long as the cosmos exists on the one hand, and their annihilation together with the cosmos on the other, is never solved, though Abravanel would probably subscribe to Maimonides' idea that the views expounded by Aristotle in his physical treatises harmonize well with the Scriptures.

[34] Feldman (1990).

[35] Ivry (2016), 90.

Seymour Feldman has acutely pointed to Abravanel's silent inheritance from Crescas and his emphasis on the gap between the reasonings we construct about the world, and God's act of creating the world out of nothing: his verdict was that Abravanel, following in Crescas' footsteps, verged on the via*negativa*.[36] This also is surely true. What I wanted to highlight in this paper is that two lines of reasoning coexist in Abravanel's philosophical theology. On the one hand, he is firmly rooted in the Aristotelian tradition as it was formed and modified in centuries of exegesis culminating in Averroes and reflected in Maimonides. On the other, if his 'Aristotelian' allegiance (broadly speaking) makes him a testimony of the long-lasting tradition of the Graeco-Arabic-Jewish philosophical tradition, and more particularly of some of its forms of thought,[37] it is this duality that adds a special interest in the study of this complex, at times eluding way to deal with Scriptural exegesis.

References

Al-Ghazālī (2000) The incoherence of the philosophers. In: Marmura ME (ed) (2000) A parallel English-Arabic text translated, introduced, and annotated. Brigham Young University, Provo, Utah

Abravanel I (2007) Letters. see below Cohen Skalli

Abravanel I (1982) Principle of Faith (Rosh Amanah), see below M. Kellner

Abravanel I (1972) Sefer Shamayim Chadashim (New Heavens), Wolf Heidenheim, Rödelheim 1828 (1972) repr. in Don Isaac Abravanel, Opera Minora. In: Jacobs L (ed) With An Introduction, vol. VI, fols 1r-51r. Gregg International Publisher, Farnborough

Abravanel I (2018) Ṣurot ha-Yesodot (Forms of the Elements), Sabbioneta 1557 repr. Warsaw 1894, and Farnborough, pp. 135–143, repr. with Italian trans. In: Coda, E (ed) (2018) Le forme degli elementi. Isaac Abravabel e la tradizione aristotelica medievale. Pisa University Press, Pisa

Cohen Skalli C (2007) Isaac Abravanel: Letters. De Gruyter, Berlin-New York

Cohen Skalli C (2009) Abravanel's commentary on the former prophets: portraits, self-portraits, and models of leadership. Jewish History 23(3):255–280

Davidson HA (1969) John Philoponus as a source of Medieval Islamic and Jewish proofs of creation. J Am Orient Soc 89(2):357–391. Doi: 10.2307/596519

Davidson HA (1987) Proofs for eternity. Creation and the existence of god in Medieval Islamic repercussions of the Kalam in Jewish Philosophy. Oxford University Press, London

Elior O (2012) Isaac Abravanel's rejection of corporeal form. Aleph 12(2):367–402

Fakhry M (2007) Islamic occasionalism: and its critique by Averroes and Aquinas Allen & Unwin, London 1958; repr. Routledge

Feldman S (1986) The end of the universe in Medieval Jewish Philosophy. Association for Jewish Studies 11(Spring):53–77

Feldman S (1990) Abravanel on Maimonides' critique of the Kalām argument for creation. Maimonidean Stud 1:5–25

Feldman S (2003) Philosophy in a time of crisis: Don Isaac Abravanel defender of the Faith. Routledge Curzon, London-New York

Gilson É (1940) The spirit of Mediaeval Philosophy (Gifford Lectures 1931–1932), Translated by A.H.C. Downes, p. 68. Sharles Scribner's, New York

---

[36] Feldman (1990), 15.

[37] On Abravanel's philsophical background see e.g. Kellner (1982a, b and 1994), Guidi (2006), Cohen Skalli (2007).

Goetschel R (1997) *Isaac Abravanel. Conseiller des princes et philosophe*. Éditions Albin Michel S.A., Paris
Goodman L (1978) Did Al-Ghazâlî Deny Causality? Stud Islam (47), 83–120. Doi: 10.2307/1595550
Guidi A (2006) Salomone come sapiente universale nel Commento al Libro dei Re, Academia. Revue de la société Marsile Ficin 8:61–80
Ivry AL (2016) Maimonides' guide of the perplexed: a philosophical guide. The University of Chicago Press, Chicago
Kellner M (1994) Isaac Abravanel on Dogma and Faith. Da'at 32–33, XLVII–LIX
Kellner M (1982) Jewish dogmatics after the Spanish expulsion: Rabbis Isaac Abravanel and Joseph Ya'bes on belief in creation as an article of Faith. Jew Q Rev 72:178–187
Kellner M (1982) Isaac Abravanel, principles of Faith, translated with an introduction and notes, Fairleigh Dickinson U.P., Rutherford
Lawee E (2001) Isaac Abravanel's stance towards tradition. Harvard U.P, Albany
Lawee E (2009) Abravanel in Italy: the critique of the kabbalist Elijah Hayyim Genazzano. Jew Hist 23:223–253
Marmura M (2005) Al-Ghazālī. In: Adamson P, Taylor RC (eds) The Cambridge companion to Arabic philosophy. Cambridge University Press, Cambridge, pp 137–154
Netanyahu B (1953) Don Isaac Abravanel, statesman and philosopher. The Jewish Publication Society of America, Philadelphia
Ogren B (2015) Ex-Nihilo creation—Isaac Abravanel on the formation of the world. In: Ogren B (ed) Evil and peace, time and eternity in Jewish Mysticism that which is before and that which is after, 83–96. Brill, Leiden
Perler D, Rudolph U (2000) Occasionalismus: theorien der Kausalität im arabisch-islamischen und im europäischen Denken. Vandenhoeck & Ruprecht, Göttingen
Seeskin K (2005) Maimonides on the origin of the world. Cambridge University Press, Cambridge
Sorabji R (ed) (1987) Philoponus and the rejection of the aristotelian science. Duckworth, London
Strauss L (2013) On Abravanel's philosophical tendency and political teaching (1937). Repr. In: Gesammelte Schriften (Edited by Meier H.), 195–231. J.B. Metzler, Stuttgart
Wolfson HA (1961) Crescas on the problem of divine attributes. Jew Q Rev 7:1–44 and 175–121
Wolfson HA (1971) Crescas' critique of aristotle. Problems of aristotle's physics in Jewish and Arabic philosophy. Harvard University Press, Cambridge, MA. 1929, repr
Wolfson HA (1976) The philosophy of the Kalam. Harvard University Press, Cambridge MA

Philosophy of Science

The Politics of Distrust

Stephen David John

Abstract This paper analyses the interpenetration of epistemic and political issues in understanding and resolving distrust in expert consensus. It is based around the case-study of debates over the safety and efficacy of the "triple vaccine" for Measles, Mumps and Rubella (MMR). The first section outlines a general model of trust in science, with a particular focus on how we learn from reports of consensus. The second section uses this model to develop an in-depth study of epistemic distrust in the scientific consensus surrounding the MMR vaccine. The final section explores the relationship between reasons for distrust and the concept of inductive risk, tentatively suggesting that arguments for deference to the consensus cannot focus only on epistemic issues but must also engage with substantive evaluative questions.

Keywords Trust · Expertise · Scientific consensus · MMR controversy · Values

1 Introduction

The vast majority of scientists agree that the triple vaccine for measles, mumps and rubella (MMR) is both effective and safe. Nevertheless, increasingly, many parents do not vaccinate their children, leading to a growth in measles outbreaks (Kmietowicz 2018). Of course, there are many explanations for the growth of "vaccine hesitancy": for example, some parents may simply be unaware of the expert consensus. It is clear, though, that some parents are not merely *ignorant* of the consensus but actively *distrust* it; although aware of the scientific consensus, they refuse to accept that the vaccine is safe. Many other cases show similar patterns of distrust in the consensus: for example, various strands of "climate scepticism" (John 2018) or debates over the safety of GM crops (Biddle 2018). In turn, these forms of distrust can have politically and socially problematic consequences: for example, an increased risk of measles epidemics or public unwillingness to accept climate mitigation policies. In such cases, we have good political reasons to care about public distrust in science.

S. D. John (✉)
Department of History and Philosophy of Science, Free School Lane, University of Cambridge, Cambridge, UK
e-mail: sdj22@cam.ac.uk

A. Fabris (ed.), *Trust*, Studies in Applied Philosophy, Epistemology and Rational Ethics 54, https://doi.org/10.1007/978-3-030-44018-3_4

How, though, do these concerns about the *effects* of distrust relate to non-experts' reasons to defer to the experts in the first place: can non-experts have political, as well as epistemic, reasons to trust the experts?

It is a commonplace that we should not engage in "wishful thinking": that the fact that believing some claim would have positive non-epistemic consequences cannot itself be a reason to believe that claim.[1] As such, the answer to my question may seem simple: experts cannot have political reasons to believe the experts. I agree that wishful thinking is irrational. However, I argue that, although political reasons cannot provide *direct* reasons to accept claims, they can legitimately play an *indirect* role in structuring epistemic practice. There is a sense in which non-experts can have political reasons to defer to the experts. I argue for these claims below. However, I am not certain of the strength of my arguments, nor of their first-order implications. The main aim of this paper is less to provide a definitive answer to my question, and more to explore the general topic of distrust in science, and how it relates to both epistemic and political concerns; even if I do not succeed in showing that non-experts do have political reasons to trust expert consensus, I hope to show that the interplay of the epistemic and political is more complex than it might seem.

Section 2 sets out a general account of epistemic trust in science, with a particular focus on the notion of scientific consensus. Section 3 uses this model to explore distrust through a case study of controversies over the MMR vaccine. Section 4 builds on work on "inductive risk" in science to develop my argument about the interrelationships between our political and epistemic practices. The conclusion discusses the limitations and implications of my arguments.

2 A Model of Trust in Science

We might say that "trust in science" has decreased; that a lack of "trust in science" creates social problems; or invite people to speak at conferences on "trust in science". The aim of this section is to clarify this concept; specifically, I argue for an "institutional" account of epistemic trust in science.

Before going on, a brief note on terminology. In the literature on trust it is customary to distinguish between "reliance" and "trust", where the latter is bound up with moral attitudes and the reactive emotions in a way in which the former is not (Hawley 2019, Koskinen 2018). While Kant's neighbours may have *relied* upon him to take a walk at the same time every day, they didn't (or, at least, had no grounds to) *trust* him to do the same; they didn't (or shouldn't) have felt betrayed when he skipped his walk to buy *Emile*. It may seem that much of what I say about trust in science is more properly thought of as reliance, because I am less concerned with intimate moral relationships than with beliefs and assumptions about how institutions incentivise performance of professional roles. My own view is that the literature on "trust" and "reliance" is misleading because it suggests a sharp distinction where

[1] For discussion of "wishful thinking", see John (Forthcoming).

there is a continuum between "mere" reliance (say, on machines) and ethically rich bonds of trust (say, between lovers). I think that "trust" in science sits along this continuum: scientists who are untrustworthy breach a professional obligation and, as such, it is (often) right to feel ethical indignation at them in a way in which we do not (or should not) feel indignant when our computers fail us. However, this is not the same kind of *personal* betrayal we feel when, say, a lover treats us badly. As such, I think that talk of trust in science, with its moral undertones makes sense. However, these issues are largely orthogonal to this paper; therefore, if the reader prefers, she should feel free to substitute "reliance" for "trust" below.

In everyday life, we use the term "trust" to refer both to expectations that others will speak truthfully ("I trust you to tell me what happened last night") and to expectations that they will perform certain actions ("I trust you to look after the plants while I'm away"). Furthermore, our everyday talk of "trust" is typically bounded. When I say that "I trust the police officer", I mean that I trust her with regard to a set of tasks; to handle the evidence carefully or to keep the peace fairly. I do not necessarily trust her to water the plants.[2] Unsurprisingly, loose talk of "trust in science" displays a similar complexity. Sometimes, when we talk about trust in science, we mean that we trust scientists to choose research topics wisely and not waste public money; sometimes we mean that we trust them to respect ethical constraints on the treatment of research subjects; sometimes we mean that we trust them to report results accurately; and so on. We can trust scientists in some of these ways and not in others: I might think that the scientist cannot be trusted to choose topics wisely, but she can be trusted to report the results of her (mis-guided) investigations accurately. My aim in this section is to investigate just one aspect of non-expert trust in science: epistemic trust; i.e. a willingness to accept others' claims about some matter in some cognitive domain. For example, I have "epistemic trust" in my wife's claims about Chinese politics, in the sense that if she tells me something about Chinese politics, then I accept this claim myself. The concept of "epistemic trust" seems central to how we think about science: the key social function of science is to be a source of epistemically trustworthy testimony.

Although narrower than "trust in science" *tout court*, the concept of "epistemic trust in science" is still ambiguous between three concepts:

a generalised trust in individual scientists (i.e. a standing willingness to accept the claims of individuals credentialed as scientists);

a willingness to defer to claims on the basis of the fact that they are the consensus within some scientific community;

a willingness to accept that scientific tools and methods are the proper ways of establishing (certain sorts of) claims.

---

[2]See (Domenicucci and Holton 2017) for some concerns about the notion that trust is a three-place relation: I suspect that Domenicucci and Holton's view can be viewed as a version of the traditional view where we specify the domain of trust in terms of a fulfilment of a role. Again, this is orthogonal to the arguments which follow.

For current purposes, I focus on the second kind of trust. Specifically, I suggest that a non-expert's willingness to defer to the consensus within a scientific community can be modelled as resting on two assumptions:

> first, a sociological assumption, that the community is organised in such a way that the best explanation for consensus on claims within the community is that those claims meet certain epistemic standards;
>
> second, an epistemological assumption, that these epistemic standards are such that the non-expert should accept the relevant claims.[3]

To spell out the relationship between the sociological and epistemological assumptions, consider two cases. First, I might concede that the community of astrologers is organised such that the best explanation of communal consensus on some astrological claim is that the claim meets the epistemic standards of astrology; nonetheless, because I think that astrological standards should not govern my own beliefs, I would refuse to accept this claim myself. Second, a climate change denialist might argue that she is willing to accept claims which have been established relative to the norms of climatology, but argue that the *actual* community of climatologists has been so corrupted by political concerns that communal consensus on claims is better explained by collusion than by the truth of those claims (John 2018).

Two features of this account of trust in expert consensus are worth noting. First, as well as providing an account of "trusting" a consensus, it provides the basis for understanding the other two dimensions of "trust" in science; a generalised trust in scientists can be understood in terms of an assumption that the community is structured such that members of the community only make claims when they are well-established relative to certain epistemic standards; the notion of trust in "the scientific method" can be understood as holding the "epistemological assumption" and/or an idealised model of how science *should* be organised, implicit in the sociological assumption. One advantage of my approach over other accounts of trust in experts is precisely that it stresses the continuity between the three usages, rather than, say, making consensus merely a reason for trusting individual experts (Goldman 2002).

Second, the account sketched above is primarily descriptive; it is a rational reconstruction of the implicit assumptions in play when we come to trust scientific claims. We can, however, use it to define two axes for assessing the normative status of an attitude of epistemic trust: first, whether a non-expert's trust would be *justified* turns on whether she can offer reasons for accepting the two assumptions which are compatible with her evidence, norms of reasoning she endorses and so on. Second, whether or not a non-expert's trust would be *warranted* turns on whether the assumptions are, in fact, true. Imagine that astrology *is* a good way of finding out about the world, but I have no evidence to believe this is so; in this scenario, trust in astrology would be *warranted*, but not *justifiable*. This is, of course, rather a far-fetched scenario, but the distinction between *justification* and *warrant* will become important later.

In this section I have sketched some issues and complexities around talk of "trust in science" and suggested a model of non-expert trust in science. Why think this model

[3]For more on this model of trust, see my (John 2018)

is accurate? The short answer, which I have argued for in more detail elsewhere, is because it provides a useful way of understanding actual debates over "trust in science" (John 2018). The next section develops this general project by showing how the model helps us understand *distrust* in vaccine safety.

3 Trust and Distrust

As noted above, there is widespread consensus among the scientific community that the "triple-vaccine" for measles, mumps and rubella (MMR) is both highly effective—at least, as long as a sufficiently high proportion of the population is vaccinated—and safe. As such, in most industrialised countries, there are MMR vaccination schemes. However, many parents do not vaccinate their children.[4] One key refrain in this phenomenon of vaccine-refusal is that parents are concerned that the vaccine might cause autism. This concern stems primarily from the work of Andrew Wakefield, a British doctor who published a paper in the *Lancet* suggesting that there *might* be a causal link between the vaccine and autism (Wakefield et al. 1998). The ins-and-outs of the subsequent scientific disputes are complex, and further complicated by their relationship to disputes in publication ethics. However, the vast majority of the scientific community think that Wakefield was wrong; there is a strong scientific consensus that the triple vaccine does not cause autism.

This section explores the phenomenon of vaccine hesitancy using the model above. To do so, it is useful to note a further important ambiguity in the language of trust. We talk a lot about "trust in science" when our real interest is in *lack* of trust: typically, we only think about trust in cases where non-experts do not defer to the expert consensus. However, as Katherine Hawley (2019) has suggested, distrust differs from a mere absence of trust. While it would be wrong for Kant's neighbours to trust him to take a walk at the same time every day, it would be wrong to say that they "distrust" him; rather, both trust and distrust would be inappropriate terms for describing this case. Similarly, when we talk about failures to defer to expert consensus we can distinguish at least three kinds of scenarios: epistemic trust without action; lack of epistemic trust; and epistemic distrust. To explain, consider the phenomenon of falling vaccination rates. One possible explanation of that phenomenon is that some parents are aware of the expert consensus and accept that the vaccine is safe and effective but refuse to vaccinate their child either on prudential or ethical grounds (say, they do not think that it is "worth" the hassle of vaccinating their child) (Dawson 2011). A second explanation points to failures on the "production" side of expert testimony; for example, problems with how scientists communicate to non-expert audiences, or problems with how the media report scientific consensus (Horton 2004). These analyses implicitly assume that parents *would* accept that the vaccine was safe if only they were aware of the true expert consensus. A third possible explanation is that

[4]For useful accounts of various aspects of the (on-going) MMR controversy, see (Boyce 2007, Fitzpatrick 2004, Horton 2004, Kmietowicz 2018).

parents are aware of the consensus, but, for whatever reason, refuse to accept that the vaccine is safe. The first case is one of "epistemic trust without action", and points us towards issues in ethics and political philosophy, rather than political philosophy. The latter two cases, by contrast, are concerned with epistemic trust, but in different ways: the second case is an instance of a "lack of epistemic trust" which could, in principle, be remedied through restructuring social-epistemic institutions; the third scenario, by contrast, involves more than a lack of trust, but active *dis*-trust.

One striking feature of much public and academic debate around vaccine hesitancy is the tendency to assimilate the first and third categories to the second: to assume that vaccine refusal must simply be a matter of ignorance of the scientific consensus. I suggest this move is problematic, because it erases the grounds for distrust in science. Specifically, I outline three reasons for distrust in vaccine safety which are offered in public discourse, and which, I argue, are not clearly unjustified.[5] Of course, you might think that these reasons, although interesting, are, ultimately wrong-headed. I return to these issues in Sect. 4 below.

First, consider the *statistically-sophisticated-anti-vaxxer*. Imagine a parent who reasons as follows: "clearly, there is still some possibility that the vaccine will have some side-effects. Therefore, when the scientists claim that the vaccine is safe, strictly they are making a claim about the *overall* likely balance of benefit and harm. Their claim is that the *average* individual is better-off taking the vaccine than not. However, my child is particularly susceptible to suffering some side effect. As such, I suspect that *he* stands to lose more than he gains from the vaccine". As Maya Goldenberg has shown, not only is the "statistically-sophisticated anti-vaxxer" an idle philosophical possibility, but it seems that many actual anti-vaxxers do often point to concerns that their children are, in some sense, special; a denial of the move from the *average* case to the *individual* case is a standard move in anti-vaccine arguments (Goldenberg 2016).

We can think of this parent as denying the "epistemological" premise: she denies that the fact that a claim has been established relative to scientific standards implies that she should, herself, accept that claim.[6] Of course, this parent could be wrong in his belief that his son is "special". Still, the general pattern of reasoning is not obviously unjustifiable: it is entirely possible that some medical intervention might be in the interests of the "average" individual in the population, but not a specific individual's interest. Indeed, a key theme in recent claims about "personalised" medicine is precisely that standard approaches to medical testing which focus on large populations

[5]What follows leans heavily on my own work (John 2011), and far richer and more recent discussion of vaccine hesitancy in the work of Maya Goldenberg (2016).

[6]Note that this case is a complex example of a far more general phenomenon: where non-experts distrust experts because the experts are addressing "the wrong question". For example, failure to defer to scientists' claims that GM crops are "safe" can be understood in terms of a concern that the scientists' technical sense of "safety" is narrower than the sense employed by non-experts (Biddle 2018). The non-experts who refuse to defer to the consensus that GM crops are safe are not denying that the scientists are right that the crops are "safe" in their narrow sense, but doubt that "being safe" in this narrow sense is sufficient grounds for believing that the crops are safe in the wide sense.

to identify an "average" effect may be problematic insofar as they overlook heterogeneity of response. Projecting claims about individual risk from population-level data is an epistemologically fraught task (Fuller and Flores 2015).

Second, consider, the *sceptic-about-consensus* anti-vaxxer. Imagine a parent who accepts the "epistemological premise"—i.e. she believes that that she should, in general, defer to claims which have been established relative to the norms of biomedical science. However, she denies that the sociological premise holds in the specific case of the MMR vaccine: she thinks that the biomedical community is not arranged such that agreement on claims comes about only when those claims are, in fact, well-established. For example, imagine a parent believes that a well-ordered scientific community is driven solely by a concern for the truth, regardless of consequences, and is characterised by a high degree of receptivity to the views of mavericks. (A view which fits well with cultural stereotypes of science). She then reads of the treatment of Wakefield, apparently drummed out of the community for his heterodox views, and notices that a key concern of many leading epidemiologists seemed to be that Wakefield's work threatened continued control of measles (a concern which is independent of the truth of Wakefield's claims). Faced with this evidence, the parent might decide that the *actual* community differs systematically from how the scientific community *ought* to be arranged, and, as such, refuses to defer to the consensus. Again, we can see this form of scepticism about the sociological structures of biomedical science as playing a role in some anti-vaxx campaigns; for example, in discussions of the treatment of Wakefield, and in discussions of the role of the pharmaceutical industry in funding research (Boyce 2007, Goldenberg 2016).

Note a complexity here: the sceptic-about-consensus might be concerned about the arrangement of the scientific community *even if* the community is, as a matter of fact, well-ordered (i.e. the fact that the claim meets the relevant epistemic standards is, in fact, the best explanation of why the claim is the subject of consensus). This can happen when the *sceptic-about-consensus* holds a "folk philosophy of science"; an account of how scientific communities *should* be well-structured, which differs from how, *in fact*, communities ought to be well-structured (John 2018). For example, the sceptic who is worried by the treatment of Wakefield holds a "folk philosophy of science" according to which well-ordered scientific communities ought to tolerate all dissent; however, some forms of dissent can be normatively inappropriate (Intermann and de Melo-Martin 2018). Therefore, her distrust is not warranted. Nonetheless, given the idealised and misleading views of science to which we are all exposed, it is easy to see how our non-expert might have formed her folk philosophy of science. It is unclear that her failure to defer to the experts is straightforwardly *unjustified*; indeed, it seems that it might even be an example of careful epistemic practice.

Third, consider the *meta-inductive* anti-vaxxer. To motivate this case, consider a possible response to folk philosophy of science distrust: that a community's track-record can give us reasons to trust, independent of any normative model of how that community should operate. I may have no clue how a bicycle workshop "should" operate, but the workshop's past record of successfully fixing my bike may give me good reasons to trust them this time around (Goldman 2002): I can have evidence that the sociological premise is true without an account of what a well-ordered community looks like. Similarly, then, we might argue that, even if non-experts have a false "folk philosophy of science", they also have strong inductive evidence that the biomedical research community should be trusted. However, the appeal to "track record" reasons for trust clearly relies on an assumption: that the experts do, in fact, have a good track record. Unfortunately, it is unclear how good the track record is in our case. There are, after all, some cases—routinely referenced by anti-vaxx communities—where the biomedical research community was wrong. In the UK, for example, vaccine scepticism is often motivated by memories of the thalidomide scandal (Boyce 2007).[7] It is not, then, contrary to the evidence to run a "meta-induction" and to conclude that there are inductive grounds to think that the sociological premise does not hold, and, as such, to withhold trust. Again, these reasons for distrust are not obviously unjustifiable.

To recap, there are two kinds of cases where parents might fail to vaccinate their children: where they accept the consensus view that the vaccine is safe and effective, but have prudential or ethical reasons against vaccination; where they do not accept the consensus view. Within the second class, we can distinguish between "lack of trust"—where parents are simply unaware of the consensus—and "distrust"—where parents are aware of the consensus, but do not accept it. Using the model of epistemic trust articulated in Sect. 1 I identified three possible grounds of "distrust" in actual vaccine debates. Before going on, note that I *do not* claim that *all* or even *most* cases of distrust in the expert consensus are actually motivated by the complex kinds of reasons above; for example, some parents may simply reject the entire intellectual structure of modern medicine; others may possess a rather inflated sense of their own ability to grasp the evidence (Sorell 2007); and so on. I focus on these three patterns of reasoning because, they reflect the reasons put forward by at least some parents and seem, to me, the most epistemologically interesting arguments from anti-vaxxers. Someone who simply rejects the entire scientific basis of vaccination seems to be making a straightforward error; someone who points to a gap between population- and individual-level claims may also be making an error, but, if so, it is a more subtle and interesting one. I hope that this exercise has illustrated Sect. 1′s general model of expertise. However, its main interest is to help us get a better grip on the practical and theoretical issues involved in thinking about trust in science. How does it do that?

4 Politics, Ethics, Vaccines

Above, I drew a distinction between *justified* trust—trust which makes sense given the reasons an agent herself possesses and recognises—and *warranted* trust, trust which is, in fact, well-placed. I think that trust in the MMR vaccine's safety is warranted. In turn, precisely because I believe that the vaccine is safe, I believe that

[7] For some more general scepticism about the predictive powers of epidemiology, see (Broadbent 2013).

others "should" believe it is safe, not only because this belief would improve their epistemic well-being, but because this belief will secure the health and well-being of their own and others' children. Understanding *why* people do not vaccinate their children is a matter of social and political importance, because ensuring that people *do* vaccinate their children is a matter of social concern. This section explores how the *political* motivation for understanding vaccine hesitancy intersects with the account of vaccine hesitancy above: are there political reasons why parents should defer to consensus?

Consider, again, the three different pathways by which some non-expert parent might not vaccinate their child: through prudential or ethical reasoning, through ignorance of the consensus, and through distrust in the consensus. As noted above, much debate over vaccination focuses on the second explanation, pinning responsibility for falling vaccination rates on reporting by the media, for example. Of course, this focus may reflect the actual situation: that many parents *are* simply ill-informed. However, it may also reflect a different issue: that responding to vaccine hesitancy seems both practically and ethically more straightforward if hesitancy is simply a function of ignorance. First, ignorance-based hesitancy is, in principle, easily remediable: we just need to make people aware of the consensus! By contrast, if refusal stems from either prudential/ethical concerns or is grounded in distrust we cannot simply *show* non-experts that there is a consensus. Rather, we must engage with and judge their *reasons* and *beliefs*; this is difficult. Second, when education measures fail, we may be tempted to respond to vaccine hesitancy through (semi-) coercive measures. It may be normatively easier to justify these measures if we frame hesitancy as an instance of hard-to-shift ignorance. Acting against someone's stated preferences is (more easily) justifiable when we think that she would consent were she better-informed than when we acknowledge that she would not even hypothetically consent to our measures. This is not to say that coercive measures can *never* be justified nor that the ethics of acting against individuals' stated preferences is simple. Rather it is to note that coercive measures seem *more easily* justified in cases of straightforward ignorance of consensus.

Unfortunately, the fact that our practical and normative situation would be simpler if vaccine hesitancy were simply a function of straightforward ignorance does not imply that vaccine hesitancy *is* a function of ignorance. (To think otherwise would be wishful thinking *avant la lettere*!) Specifically, I have suggested that vaccine hesitancy is often grounded in reasons for distrust, rather than mere ignorance, and that some of these reasons are epistemologically complex. How, then, does my analysis affect both the practical and normative issues concerning the proper response to hesitancy? The previous paragraph suggests one way into these questions: although distrust is clearly more complex than mere lack-of-trust, one might still think that, ultimately, it is unjustifiable. As such, although strategies for tackling distrust will have to be more complex than strategies for tackling ignorance, the two situations raise roughly similar issues: preferably, we should identify education strategies to promote proper thinking; if these fail, then more coercive strategies may be justifiable on grounds of hypothetical consent.

These thoughts may be particularly appealing in the cases outlined above. It might seem that, even if we can reconstruct distrust as involving epistemically complex reasons, ultimately, these pathways to distrust are epistemically problematic as they involve peculiar weightings of evidence. For example, even if we concede a "gap" between population-level data and individual-level claims, it seems that the population-level data makes it exceedingly unlikely that the vaccine is unsafe for particular individuals; similarly, although it is *possible* that the consensus is the result of conspiracy, rather than genuine epistemic agreement, it seems highly implausible that this is so; although the track-record of bio-medicine is not spotless, a few high-profile mistakes seem a weak evidential basis for thinking that biomedical communities are systematically flawed. To use the distinction above, we might suspect that anti-vaxxers' distrust in science is not only *unwarranted* but *unjustified*: i.e. even assessed relative to their own beliefs and commitments, their distrust does not stand up to epistemic scrutiny. As such, they may be, in principle, amenable to education, and, if this fails, in principle, subject to coercion on grounds of hypothetical consent.

Unfortunately, even if distrust involves an apparently odd weighting of the evidence, this does not suffice to show that it is therefore unjustified. To see why, consider a commonplace in recent philosophy of science that scientists take inductive risks: i.e. that in choosing to accept or reject claims they must balance the risk of false positives against false negatives. Some have argued that, insofar as there is no purely epistemic way in which to judge the relative costs of these different sorts of error, scientists must, therefore, appeal to non-epistemic values in their practice (Douglas 2000, John Forthcoming, Rudner 1953, Wilholt 2009). Let me zoom out from the details of these debates to note a far broader point they illuminate: that some claim is "well-established" in some circumstances does not imply that it is "well-established" in all circumstances. Rather, our willingness to act on claims which might be wrong does (and should) vary with the non-epistemic costs of different sorts of error. In general, the greater the costs of acting on a false positive, the more evidence we should demand before accepting some claim. This is not to say that our non-epistemic values can give us *direct* reasons to accept or reject claims—our non-epistemic values cannot themselves justify us in believing ill-supported claims—but it is to say that they can play an indirect role, by setting the "degree of certainty" which some proposition must enjoy before we properly accept it. Our non-epistemic values might (indirectly) give us reasons to weigh our evidence in apparently irrational ways.

Consider, then, the vaccine case. Imagine a parent who accepts that claims about vaccine safety and efficacy are well-established, relative to the epistemic risks which scientists do (or should) be willing to tolerate. However, she also reasons that the stakes in her situation—where her own child's health is at risk—are such that she should demand *higher* standards than govern the scientists' acceptance and assertion of claims; as such, she refuses to accept the scientifically well-established claim. Such refusal to defer to the experts is not necessarily epistemically irrational, insofar as the parent does not deny either the evidence or the consensus. Rather, all she does is to insist that there is a difference between "scientific certainty" and "practical certainty" (John 2011). This is, of course, a very abstract way of thinking about things. I suggest, however, that this pattern of reasoning is not obviously epistemically unjustifiable.

Furthermore, we can interpret the pathways to distrust outlined in the previous section as justified along broadly similar grounds. For example, we should not understand the parent who distrusts the consensus on the basis of past errors in the biomedical community as thinking it is more likely than not that the community is flawed: that would be a straightforward epistemic error. Rather, we can see her as agreeing that it is *very likely* that the community is well-ordered refusing to accept that a *very likely* claim is *likely enough* for her to accept it, given the non-epistemic costs associated with accepting that claim if it is, in fact, false—exposure of her child to a dangerous vaccine.

If non-epistemic values can play a justifiable, indirect role in decisions about acceptance, and, hence, in decisions about trusting or distrusting the expert consensus, we have two options. The first is to treat vaccine hesitancy as essentially beyond rational resolution: we might think of the clash between pro- and anti-vaccination camps as akin to a clash between essentially incommensurable worldviews. The second option is to engage with, assess and criticise the non-epistemic values underpinning or expressed in unwillingness to defer to expert testimony, in the hope that we might either convince anti-vaxxers to change their values, or, at least, find justifications for coercing them to vaccinate their children.[8] Given the norms of liberal democracy, we should explore the second option: how, though, might we criticise the non-epistemic values underlying expert distrust, given that at least one of the operative values—concern for one's child—seems both natural and admirable?

Our best hope is, I suggest, to exploit a structural analogy between the case of "high-standards" distrust and the classic free-rider problem in political philosophy and bio-ethics. To explain: much philosophical literature on vaccination is concerned with the scenario of a parent who believes that the vaccine is safe and effective but fails to vaccinate his own child as the result of a rational calculation that, as long as everyone else vaccinates *their* children, vaccinating *his* child is pointless, because "herd immunity" will obtain regardless of his actions (Dawson 2011). From a purely prudential perspective, there is no point taking a day off work to vaccinate one's child if your child will enjoy the benefits of others being vaccinated regardless of what you do. Such behaviour presents a classic philosophical puzzle: taken in isolation, the free-rider's actions *harm* no-one, but they still seem *wrongful*. One way of spelling out this concern is that the behaviour is problematic because it violates obligations of fairness: one simultaneously enjoys a public good that is generated by everyone else in the community undertaking some action with some cost while oneself refusing to that action.

In principle, we could transpose these concerns to the case of distrust in experts: that is to say, if we could show that widespread commitment to trust experts generates certain sorts of public goods, and that sceptical parents rely on these public goods in adopting their "very high epistemic standards", then we could treat adoption of high standards as a form of free-riding. This is, of course, a sketch of an argument: to fill it out, we need to show that widespread agreement to "normal" standards does generate

[8]This framing is heavily influenced by Martin Kusch's excellent account of the political philosophy of risk (Kusch 2007).

goods which parents rely upon in raising their own standards. Is this plausible? Here is one possibility: in general, individual planning is made more predictable and reliable when we can each assume that, in our community, there is shared, widespread commitment to certain sorts of epistemic standards. Very roughly, when we know that everyone else will treat certain sorts of claims as established for all practical purposes, then we can better predict how each will respond to shared aspects of the social-epistemic environment. If so, we can understand a widespread practice of deference to scientific experts—a practice which is ultimately grounded in epistemic considerations—as generating a political reason to defer to experts: that one enjoys the benefits of co-ordination which follow from everyone else deferring to experts. What is wrong with "very high standards" distrust is not so much that it is epistemically flawed, but that it conflicts with the demands of fairness: one is simultaneously making predictions about the social world which assume that everyone else will continue to use normal standards while, oneself, using higher standards. Distrust may be epistemically rational, but politically unfair.[9]

5 Conclusion

My argument for a fairness-based obligation to defer to the expert consensus clearly requires refinement; as such, it is unclear that it is successful. Even if it is successful, presumably all it generates is a *prima facie* obligation to defer, to be balanced against questions about freedom of thought, parental prerogatives, and so on. Furthermore, even if there is a fairness-based obligation to defer to the expert consensus, it is not clear what follows for the first-order question of how best to respond to actual vaccine hesitancy. Still, there is something valuable in pursuing these thoughts. As noted above, typically we care about trust and distrust in science for political reasons. If non-experts do not trust the consensus among the physicists at CERN, this is a shame for them—they may be missing out on epistemic enlightenment—but it does not seem an issue of pressing practical importance. By contrast, when non-experts do not trust the consensus among the biomedical scientists on vaccine safety, this is not only a shame, but a political problem. However, it is tempting to think that, although our reasons to care about distrust are *political*, epistemic distrust is, itself, a straightforwardly *epistemic* problem, in the sense that it reflects some ignorance or flawed reasoning. By means of a model of trust in experts, a case-study and arguments in philosophy of science, I have argued that we should not draw a strong distinction between *epistemic* and *political* reasons for distrust. Once we recognise that non-epistemic values can play a legitimate, indirect role in how non-experts respond to expert consensus, then we must either accept that there can be cases of irreconcilable disagreement or we can look for good non-epistemic reasons to combat scepticism.

[9]See my (John 2011) for a slightly different version of the same argument—the current argument builds on that argument in relating it far more closely to notions of trust and consensus.

If we want good arguments for trust in science, we need to rethink the distinction between the epistemic and the political.

References

Biddle JB (2018) "Antiscience Zealotry"? values, epistemic risk, and the GMO debate. Philos Sci 85(3):360–379

Boyce T (2007) Health, risk and news: the MMR vaccine and the media. Peter Lang, New York

Broadbent A (2013) Philosophy of epidemiology. Palgrave MacMillan, London

Dawson A (2011) Vaccination ethics. In: Dawson A (ed) Public health ethics: key concepts and issues in policy and practice. Oxford University Press, Oxford, pp 143–153

Domenicucci J, Holton R (2017) Trust as a two-place relation. In Faulkener P, Simpson T (eds) The philosophy of trust. Oxford University Press, Oxford

Douglas H (2000) Inductive risk and values in science. Philos Sci 67(4):559–579

Fitzpatrick M (2004) MMR and Autism. Routledge, London

Fuller J, Flores L (2015) The risk GP model: the standard model of prediction in medicine. Stud Hist Philos Sci Part C: Stud Hist Philos Biol Biomed Sci 54:49–61

Goldenberg M (2016) Public misunderstanding of science? Reframing the problem of vaccine hesitancy. Perspect Sci 24(5):552–581

Goldman A (2002) Experts: which ones should you trust? In Goldman A (ed) Pathways to knowledge. Oxford University Press, Oxford

Hawley K (2019) How to be trustworthy. Oxford University Press, Oxford

Horton R (2004) MMR: science and fiction. Granta, London

Intemann K, de Melo-Martin I (2018) The fight against doubt. Oxford University Press, New York, NY

John S (2011) Expert testimony and epistemological free-riding. Philos Q 61(244):495–517

John S (2018) Epistemic trust and the ethics of science communication: against transparency, openness, sincerity and honesty. Soc Epistem 32(2):75–87

John S (Forthcoming) Science, truth and dictatorship: wishful thinking or wishful speaking? In *Studies in History and Philosophy of Science Part A*

Kmietowicz Z (2018) Measles: Europe sees record number of cases and 37 deaths so far this year *BMJ* 2018;362:k3596

Koskinen I (2018) Defending a risk account of scientific objectivity. Br J Philos Science

Kusch M (2007) Towards a political philosophy of risk. In Lewens T (ed) Risk: philosophical perspectives. Routledge, London

Rudner R (1953) The scientist qua scientist makes value judgements. Philos Sci 20(1)

Sorell T (2007) Parental choice and expert knowledge in the debate about MMR and autism. In Dawson A, Verweij M (eds) Ethics, prevention and public health. Oxford University Press, Oxford

Wakefield AJ et al (1998) Ileal-lymphoid-nodular hyperplasia, non-specific colitis, and pervasive developmental disorder in children. Lancet 351(1998):637–641

Wilholt T (2009) Bias and values in scientific research. Stud Hist Philos Sci Part A 40(1):92–101

Information, Science and Democracy, for an Ethics of Scientific Knowledge

Giuseppe Longo

Abstract The modern theories of information opened to way to extraordinary new tools for communication and knowledge construction. When identified instead with the world (DNA is a complete information carrier and a program for ontogenesis, atomic particles and planets exchange information ...), a new form of scientism affects knowledge and trust in science. Vague and implicit assumptions, with no precise references to the sciences of information, force strong consequences, in biology in particular. Similarly, the powerful optimization methods that make intelligible the dynamics of inert matter are expected or used to direct also human activities and replace human government by automatic governance. We are then meant to follow the formal, mechanical rule or the unique optimal path, instead of inventing new forms of sense in human interactions. Yet another fantastic tool, the newly available Big Data, is being considered to substitute scientific invention by correlations among numbers: fortunately, mathematics a priori shows that a deluge of spurious correlations may be found in any very large set of numbers. An ethics of knowledge may then be proposed in order to reconstruct the trust in science, as critical thinking at the heart of democracy.

1 The Sciences of Information

The concept of information has been dealt with by at least two rigorous and important scientific theories: elaboration of information, after Alan Turing, and communication or transmission of information, after Claude Shannon. These two theories are based on fundamental properties of mathematical invariance, which is to say these authors

Extended and revised synthesis of an interview by P. Bartolini in Italian (ROARS, September 2016), translated into English by M.R. Doyle and S. Savic.

G. Longo (✉)
Centre Cavaillès, CNRS and Ecole Normale Supérieure, Paris, France
e-mail: giuseppe.longo@ens.fr
URL: http://www.di.ens.fr/users/longo

School of Medicine, Tufts University, Boston, US

A. Fabris (ed.), *Trust*, Studies in Applied Philosophy, Epistemology and Rational Ethics 54, https://doi.org/10.1007/978-3-030-44018-3_5

singled out some key mathematical notions or properties that can be transferred from one context to another, while preserving that which matters. In both cases, typically, the characteristics of information do not depend neither on the code chosen (apart from the negligible price of coding, be it by 0 s and 1 s or by numbers from 0 to 9 or any other sequence of signs) nor on the material support: one can elaborate information on transistors, chips, silicon, and can transmit signals by wires, drums, smoke .... This ancient invention, formalized in a revolutionary manner by Turing in 1936 and by Shannon in 1948, has enabled us to distinguish between *software* and *hardware* and to propose an autonomous theory of programming, i.e. of elaboration of information, or of transmission of information (Shannon)—both independent of the specific coding and of the material support (the hardware), a fantastic theoretical and practical feat. In Manchester, in 2012, at a conference marking 100 years from the birth of Turing, the students constructed a Turing Machine (1936), the mathematical prototype of modern computers ... out of Legos. This machine worked, although of course a little bit slowly.

More precisely, in the 1930s, Gödel, Church, Turing ... each of these mathematicians gave his own, very different definition of the notion of "computable function"—these turned out to be all equivalent. In short, a function is computable when it can be described by an algorithm—a set of instruction or formal rules in a sufficiently powerful language. The consequence of that non-obvious equivalence results is that this notion does not even depend on a formal language or system of calculation used. Independence or invariance of a notion, typically from the formal system used, the coding and the material support, it is the force and the scientific relevance, in particular, of the notions of information and program.

2 Mysticism and Scientism

As it often happens, the mystics have taken the beautiful mathematical invariance in the two theories of information as an absolute and have applied it to all phenomena, including natural ones. Thus, when one of the leaders of the Human Genome Project writes that, soon, we will be able to transfer genetic information on a compact disk and claim/assert "Here is a human being, this is me!" (Gilbert 1992), he applies the distinction between software and hardware, fundamental in computer sciences but catastrophic when it is about understanding the living. Gilbert's statement, a rather common attitude in molecular biology, betrays an alphanumerical, linguistic vision of biology as just an analysis of software, independent of coding, of its material realization and, of course, even physical dimensions—in discrete data types, phenomena in any finite physical dimension be encoded in one dimension, a string of 0s and 1s. This "linguistic" approach is very well described by François Jacob and Jacques Monod, two great scientists, protagonists of molecular biology, rare among those who have made explicit a broad theoretical reflection on the subject: "The surprise is that genetic specificity is written out, not with ideograms like in Chinese, but with an alphabet" (F. Jacob, opening lecture, Collège de France, May 1965).

All this would reassure us that God is on our side, as confirms Francis Collins, director of the National Human Genome Research Institute, in June 2000, publicly announcing the decoding of five human genomes together with Venter, Clinton and Blair: "we have caught the first glimpse of our own instruction book, previously known only to God". Mysticism and theology, or researchers' enthusiasm about an extraordinary technical performance? Which philosophy and which practice of biology are they proposing? Which vision of the human?

It is firstly the sense of radical materiality of life that one loses, that within which one cannot distinguish the software and the hardware: The living state of matter is made of this DNA, RNA, only, of these membranes and their physics-chemistry and of nothing else …. Synthetic biology in fact transfers fragments of these from one to the other: The DNA or fragments of DNA in a membrane taken from another cell; it does not re-write neither transfer software onto another hardware, a practice common in computer science. When the colleagues from molecular biology will know how to transfer information and biological programs in a DNA on Lego and make a cell that works—then we will be able to affirm that they have finally "extracted" and separated information and instructions, the software of the living, written by God or not, from its hardware, in the mathematical and informational sense of the words. Are these only innocent metaphors? No: The characteristic of a metaphor is that it transfers and enriches the meaning of a theoretical proposition. In our case, it is a reference to commonsensical, vague, poorly defined notions of information and program (where is the operating system, the compiler in a cell? Which are the proper invariant properties of computing and information mentioned above?), but from which strong consequences have been derived, heavy weight onto knowledge, as it may be shown in reference to cancer research (Longo 2018b). This is unacceptable in science, because this imposes a bias without rigorously specifying its origin and forces research onto tracks whose principles remain implicit, a form of "scientism" that affects science and trust in science. For example, as molecules, DNA in particular, are the obvious locus for discrete/digital coding of information, it follows that "everything is molecular", or even "everything is genetic", the most organized chemical structure in an organism. Thus, the claim that "any phenotype has a causal origin in the genotype", including, say, the gene for monogamous behavior (Nature 1999), as, for many, genes should broadly encode also behavioral information (Plomin 2018).[1] Instead, DNA is the extraordinary physico-chemical trace of Evolution, a fundamental *constraint* in onto-phylogenesis, as I will hint below.

[1]Unfortunately, these simplistic views of genetic determination reach the general public: "Being Rich and Successful Is in Your DNA" (*Guardian*, July 12, 2018); "A New Genetic Test Could Help Determine Children's Success" (*Newsweek*, July 10, 2018); "Our 'Fortunetelling Genes' make us" (*Wall Street Journal*, Nov. 16, 2018).

3 Strong Consequences from Vague Hypotheses

As for an analogy, the genocentric/informational approach in biology strictly parallels the geocentric approach to the Universe. They are both based on commons sense: Everybody knows what information and programs are, as it is evident that the Earth does not move … Then, from both vague "evidences", complicated theories and strong consequences have been derived, based on ad hoc epicycles on top of epicycles for every planet, genes controlling gene expression for every phenotype. Science instead is made by putting forward strong and rigorous hypotheses, which go against common sense, against the idea that the Sun raises on motionless Earth, that light travels along a Euclidian line, or that a measurement remains the same if one changes the order between position and momentum (the non-commutativity of quantum physics) …. From vague hypothesis on information and programs, which have for too long governed molecular biology, an understanding of how transmission and elaboration of information may work on macromolecules has been *deduced*. That is, the strong idea of the "*Necessarily stereospecific* molecular interactions explain the structure of the code … a Boolean algebra, like in computers" (Monod 1970) was derived. This prevented observing the stochastic character of these interactions, already known in physics-chemistry, whose probability depends on the context—the cell, the organism, the ecosystem—for a classic reference, see (Elowitz 2002) and for more references (Bravi and Longo 2015). In this way, little attention has been payed to the possible role of endocrine disruptors of finite combinations of the 82,000 molecules that we have produced and thrown into the ecosystem in a little more than a century (Diamanti-Kandarakis et al. 2009), as most of them are far from presenting exact affinities with bio-molecules. Yet, they do affect hormonal cascades as they modify the *probabilities* of interactions, in particular with cellular receptors. Or, they may even replace hormones, in low, but sufficient, probabilities due to lower chemical affinities or by changes of affinity constants due to modifications of the chemical context, (Vandenberg et al. 2012). Of course, stochastic macromolecular interactions and stochastic gene expression, in a cell, a tissue, an organism constraining those interactions, are incompatible with the current notions of information and program, their determinism and their focus on DNA, elaborating and transmitting information in the exact Boolean fashion we find in computers.

In this way, by affirming that "the organisms are simple vehicles of genetic information" (to take a 2002 reference text in molecular biology of evolution) in the service of the "selfish gene" (as in a popular philosophy), it has been believed, for example, that we could completely pilot the development of plants by genetically modifying them (GMOs). To be more precise, GMOs are the children of the "central dogma" of molecular biology (Crick 1958), according to which information propagates from DNA to RNA to proteins in a one-way and linear manner (except for some retroactions RNA-DNA). To the contrary, today, we understand the strong interactive and retroactive eco-systemic effects, acting on a number of levels. For example, GMOs heavily affect the microbiome and roots' mushrooms, essential symbionts of plants. As a consequence, they gradually transform humus into a sort of sterile sand (Bizzarri 2012). Again, the exact macromolecular interactions (stereo-specificity, the key-lock paradigm) were *derived* from the informational/programming myth, which justified or gave a pseudo-rigorous frame to the genocentered approach, against empirical evidence. And this is massively affecting our lives and, in many, the general trust in science.

When we speak of the consequences of this "programmatic/informational" and DNA-centered vision in cancer research, the organismal perspective in (Sonnenschein and Soto 1999) may be an alternative reference. It was confronted by harsh criticisms for at least a decade. But a paradigm shift is taking place (Baker 2014). Recent articles by leading researchers in the informational perspective acknowledge the failure of the search for the "signal" from the carcinogen "de-programming" the genetic program. So, R. Weinberg, a biologist from the MIT, is among those working for long on search for genetic therapies for reprogramming DNA, promises made over and over since 1971, the beginning of "War on Cancer" by Nixon, and still today. In (Weinberg 2014), he acknowledges the failure of the programming/informational, genocentered approach to cancer. Similarly, Gatenby, another leading personality in the field, in a 2017 paper in *Radiology,* acknowledges the little or absence of *causal* relation between DNA mutations and the various forms of cancer, see (Longo 2018b) for this and more references. Where has information gone? Should it be just diluted in epigenetics? Ok, then one should define the invariant properties that make the sciences of information robust and that should be used in an epigenetic frame: is the proteome and its interactions that support the biologically relevant information? Or use another word and concept or another *explicit* theory.

But, careful: Molecular biology has given us an immense quantity of data and has shown evidence of fundamental mechanisms. It is extraordinary to see the experimental fineness and originality of methods that enabled Changeux, Jacob, Monod and Lwoff to discover the mechanism of chemical allostery and of the lactose operon, a regulation of the genetic expression by DNA itself. However, on the basis of the operon specific case, they leave aside the epigenetics, to say the least: genes regulate gene expressions …. So, they do not cite Barbara McClintock, the first author to analyze the *epigenetic* regulation of gene expression. In fact, since the 1950s, the studies of major geneticists like McClintock and Waddington, suspected of "Lamarckism" (by giving a role to external influences on DNA), have been abandoned, for decades. They had always studied the chromosomes, an extremely important component of the cell, in a context: the proteome, the cell, the organism. Evelyn Fox-Keller tells this story in a book on Barbara McClintock. In this way, for decades, molecular biologists have researched regulators of genetic expression in DNA, assigning it the role of the complete informational description of an organism, like an Aristotelian *humunculus*, very small indeed, but completely codified in chromosomes and therefore "programmed" (yes, we are modern). On the contrary—some geneticists have known this for 60 years—every relevant molecular cascade, from DNA to RNA, to proteins, depend causally on the context and on its material, physico-chemical, realization. The activity of these macromolecules is context driven or used by the context—the cell, the tissue, the organism—in its irreducible tri-dimensionalty and materiality: No software as a mathematical invariant, no information as independent of coding, hardware, dimensions etc. And since several years we know that the individual cells, taken in small groups from cancerous tissues and transferred into healthy tissues take up normal function, without necessarily "reprogramming" their DNA.

4 Randomness and Bio-Resonance

The concept of randomness in science is a delicate one. It is not an absolute. It depends on theories. It should be understood as the unpredictability with reference to the intended theory (Calude and Longo 2016a, b): it is what the theory shows/assumes to be unpredictable. In physics, classical and quantum dynamics propose two concepts of randomness, epistemologically and mathematically distinct. The first depends upon the joint role of non-linearity and of the physical measurement, namely upon the interactions in a dynamic and upon the access to the world given by measurement. A fluctuation/perturbation below measurement is the hidden cause of a phenomenon observable and measurable after some time—non-linearity may amplify the hidden fluctuation (Poincaré wrote in 1902: "and we then have a random phenomenon"). It is about a chaotic determinism of throwing dice, of a double pendulum or the solar system (luckily for us, in quite different temporal scales!). Quantum mechanics proposes another definition, which is intrinsic to the theory and starts from the indetermination (and non-commutativity) of measurement. Moreover, the phenomenon of "entanglement" distinguishes quantum randomness, mathematically, from the classical one. In biology, or at least in the cell, these two forms of randomness overlay each other. There is significant evidence of quantum phenomena with phenotypic effects (Buiatti and Longo 2013) and these may be amplified or dumped, according to the context, by classical phenomena. Moreover, the interaction of different levels of organization (molecules, tissues, organism) produces effects of bio-resonance, as we defined it, which may destabilize and stabilize an organism. But the most important thing is that this randomness, in biology, is not "noise" as in (Monod 1970), a notion related to information, and like many claim still today (Bravi and Longo 2015). It is instead an essential component of the variability, adaptability and diversity of the living and thus of its structural stability on the level of species, of populations and also in an organisms. They are all "more stable" when adaptive and diverse and this is partly due to the peculiar form of randomness one has to describe in biology, as we hinted above (for more about unpredictability in Evolution, see (Longo 2016a, b)).

To take an example, the immune system is a paradigm for the "functional generation" of randomness, essential to the stability and adaptability of an organism. But even an organ of such a uniform appearance such as the liver contains up to 50% of cells with a "wrong" number of chromosomes or with all kinds of mutations. By the diversity of enzymes' production, this contributes to the resilience of the organ (and the organism) to a variety of unforeseeable toxic shocks. These biological features have nothing to do with "noise" which destroys information. Instead, adaptability, exploration of diversity, changing niche constructions ... are at the core of biological dynamics. They simultaneously modify organisms and ecosystems, phenotypes and functions, and therefore the "observables" themselves in the evolutionary processes. Following a fundamental chapter of Darwin's Origin of Species, organism's "correlated variations" and its "extreme sensitivity" to internal and external conditions make it, but also the ecosystem, a conceptual challenge that cannot be envisaged merely as elementary "stacking" of elements. Machines are constructed by stacking constitutive elements, ones on top of others: We construct them by association, assemblage of components, nuts and bolts, chips and bits, simple and elementary. Occasionally, some noise may affect the construction or its functioning.

Also, computers and networks of computers are superpositions and a very complicated blend of simple and elementary components, which elaborate and transmit *bits* of information. In the natural sciences, on the contrary, it is not necessarily the case that the fundamental is *elementary*, nor that elements are *simple*. Galileo's or Einstein's theories are fundamental but do not concern the elementary, such as Democritus's atoms or quanta. Moreover, these last ones, as much as the cells that compose the elementary of the living, are not at all simple. But, even more crucially, a multi-cell organism is not made by adding tissue to it, by attaching organs to it, putting eyes in holes ... like in all machines. On the contrary, it is made by differentiation from one cell, a zygote, which is already an organism and which, at each cellular differentiation, keeps its organismal unity. There, we have an original complex object, a cell, which further complexifies by constructing a new unity by differentiation, and not a stacking of elementary and simple bricks, glued together. All this occurs in a permanent construction of diversity, radically unpredictable, during phylogenesis: organisms co-construct niches in changing ecosystems that "enable" them (Longo et al. 2012). The changing space of possibilities singles-out biological evolution from physical dynamics (Longo 2016a, b).

5 Information Everywhere, an Abuse

Besides the reinforcement of reductionists approaches in biology due to the notion of digital, programmed/transmitted information, I do not see how that sort of analyses can be improved by saying that "information is everywhere". Which are the principles of this extended and even more vague notion? Is this a reference to the remarkable new field of "Geometry of information", a new mathematical area rigorously dealing with homotopy theory and continuous symmetries and their breaking (Barbaresco and Mohammad-Djafari 2015)? Instead, in an attempt to make the commonsensical reference to digital information more rigorous, Maynard-Smith, in a 1999 article we discuss in Perret-Longo's paper in (Soto et al. 2016), confuses, in the examples, Turing-Kolmogorof and Shannon-Brillouin dual approaches to entropy and complexity as amount of information on discrete data types—in one case they are covariant in the other contra-variant. In (Longo and Seno 2018), (Longo 2019) and (Montévil 2020), we tried to be more specific as for the use of delicate concepts such as negentropy, anti-entropy etc.

More generally, we do use, as humans, *gradients* (variations) of energy or matter to transmit and elaborate information (modulation of frequency, of electric flows, drums, smoke ... switches). Gradients of flows, in organismal and cellular exchanges typically, are very important. The claim that one has information, in a natural phenomena, any time a gradient matters more than the *quantity* of a flow, is an amazing abuse. It is an anthropomorphic-magic projection on nature and it blurs more appropriate scientific analyses of the role and nature of those gradients, those flows and their specific material structure. "Information" is too important today, to be left to a vague reference to an un-specified notion and then diluted on nature: robust, well adjusted theories to empirical evidence should be given, in search for bridges, dualities, unifications .... These is how the remarkable theories of information mentioned above have been constructed, their relations are still to be fully developed. Physics, since Galileo's inertia (momentum conservation) keeps proposing new observables and pertinent parameters: Force, energy, entropy ... new quantum observables. A change of scale forces a change of theory: Quanta, thermodynamics, hydrodynamics, Relativity ... all *incompatible* theories. Unity is very difficult to achieve: Newton, Maxwell, Boltzmann and Einstein and no others proposed remarkable unifications—each a scientific revolution. A book by three physicists who work on the "borderline" of theories explains this beautifully, a review is in (Longo 2016a). In which way, claiming that both atoms and planets exchange information would help in unifying the quantum and relativistic fields, an open issue since a century? Diversity of tools and concepts, presented with rigor and facing the variety of phenomena (and scales!), is at the core of the scientific knowledge we may trust: The endeavor towards unity is a further, very difficult achievement. Scientism, against the history and the practice of science, let many believe that science is the progressive "occupation" of reality with familiar tools and concepts, possibly just one—information for example. This is a way to claim the full control on nature, by handling information, like the information in the DNA of plants (the GMOs) or the incredible fuzz made on "genome editing". A fantastic discovery, the way some viruses modify DNA, has been transformed in financial support for actions on DNA, which result in plenty of uncontrolled side-effects (Sikandar 2019). These explorations can be a component of scientific work, what is wrong is its mythical handling and the resulting distortion of research and its funding by phantasmagoric promises, see (Audétat 2015) for a survey. Finally, this hides a mechanistic view of nature, which should be controlled like our piecewise constructed machines, with no appreciation of the peculiar organismal unity and historical dynamics in animal, plants and ecosystems.

6 Optimal Governance Versus Government

Besides "mechanisms", scientism also claims to understand and to govern the world by optimization methods. For example, the notion of "Human Resources", which analyzes work in the same way as it analyzes material resources, have been invented in the ex-USSR. Its aim was the optimization of all components of production, including human work, by the mathematical methods of "linear programming". This is how the modern idea of purely technical government is born, put in the hands of engineers of the soviet Gosplan; this is what we call, today, "governance". These methods have since found their place in our major private companies in the West and, then, also in the public and national sectors, from health to government. Shouldn't we have to govern them like we do govern (big) business? The government, should it be entrusted to the "best" entrepreneurs, like it was in Berlusconi's Italy, with a party as centralized and authoritarian as the worst Stalinist party of our democracies. Trump is a further example of this identification: Business governance rules government. This is how the circle closes. In the State Enterprise, Stalinist or neoliberal, scientism and governance blend wonderfully.

Alain Supiot in his book *Governance by Numbers* (2017) more closely explains the difference between "governance" and government by law. The former is about an "objective" management, according to formal rules, potentially mechanizable, independent of all contexts. It uses methods of optimization, like in the "linear programming techniques" invented in URSS, yielding one single possible and optimal path, a "geodesic". Geodesics are produced also by the laws as equations of Walras equilibrium economy: These spontaneously determine a unique possible paths, an optimal one—societies have just to follow it. To the contrary, the law of man, the "government", is motivated by its social or human meaning, when first discussed in the agora, in the moment of its being voted, and afterwards, when it is *interpreted* by a government or a judge who apply it in their domains, who give it a possibly new contextual sense.

In fact, the term "natural law" has been historically articulated into the "law of man" (and of the gods) in a very interesting way. In different cultures, it is interpreted in very different ways (Longo 2017). The formal rule, based on numeric or formal writing, is instead manageable by automatisms independent of interpretative ambiguities, of which we make in the best case a "fine tuning". Governance may be largely implemented by machines elaborating information; it is claimed to be "objective". In the Merkel-Sarkozy agreements, for example, it is written that the punishment for the States that transgress the rules about the deficit will be implemented in an "automatic way", potentially mechanizable. But the essential question is that of democracy, that of law which has to be discussed and interpreted, which gives meaning to the life in a community, with the possibility of a real and organized disagreement: The analysis of Stalinism and of neo-liberalism, European or American is subordinated to it. So, similarly to the huge Soviet industrial conglomerates, Americans corporations, since the Securities and Trust Act of 1933, have the status of a medieval fiefdom or of a legal person, a State within the State.

When we talk about science, within large, once enlightened corporations, like in renaissance principalities, such as IBM, ATT-Bell Labs or Digital, with which I collaborated at the end of the 1980s, there were spaces of extraordinary freedom for research: Large groups of very diverse researchers could think freely, protected by the master/corporation, thanks to a long term, enlightened vision of industrial research. Apple, Google and Microsoft offer still today remnants of this, autonomous research centers with about a few dozens people free to think, but these are no longer hosts of hundreds of researchers who, until the beginning of the 1990s, explored all possible directions of American industry: research for immediate profit dominates. Apple and Google have begun with great innovative ideas, at the end of 70s and 80s, but today they mostly apply, in computer memories all more miniaturized, powerful and faster, about the same ideas—or transfer and develop new ideas coming from public research, like the new Artificial Intelligence (the so called Deep Learning, a major technical innovation). When we talk of Microsoft, Roberto di Cosmo, professor of computer science in Paris, explains its growth in *Planetary Hold-up*, as a moment of disruption without research content in the 1980s. More generally, the significant reduction in industrial research in the past 20 years has been described by a number of authors, see (Seno 2014), (Rafols et al. 2014), (Audétat 2015) and many others. It is one of the main consequences of the transformation of industrial property into a shareholders scheme, managed by universal managers, interchangeable, who have only to increase the value of a company's shareholders in a short or very short period of time. The permanent hype for new gadgets is transforming the public's trust in science: It is an instrument for advertising more than for knowledge construction. In this frame, note that the value of shares, in view of volatility and the evaluation of risks, is a matter of information, including on insurance assets, and on insurance based on information, far away from actual products and productivity. The continual flow of announcements of new technological applications manipulates this information and the associated shares values: It is at the heart stock market activities. This is the cultural hegemony that, in my view, dictates the universal role of the notion of information. No scientific argument nor evidence may be strong enough to oppose to a hegemony which is first economic and political.

As a matter of fact, all of this is of little help in facing the problems of our world's complexity: This requires first discussing the method and processes of research. Any research activity has its own method and concepts to be spelled out explicitly, as hinted above as for the notion of information, possibly on the grounds of an *ethics*: questioning one's own principles and motivations of knowledge, by making a "step aside", look at them at a distance. This requires giving oneself time to think, without knowing exactly where one is going, open to being judged, and severely so, in order to be hired, promoted, closely evaluated on the basis of one's *past* creative work, beginning with a PhD. Thesis work and so on, say, and not on promises. The computational governance of research by numbers, by "bibliometrics" typically, does not work either, nor financing (only) huge projects. Bibliometrics reinforces dominant fashions, kills diversity and critical thinking, an essential component of science and … democracy. The identification of democracy with the majority vote, disregarding the division of powers and the formation of alternative views, is a major danger in all our countries. It similarly endangers science by the "vote of the majority" of all colleagues on Earth, counted by the number of quotations, by machines searches on Internet, in the short term (the impact factor concerns at latest two or five years from publication, which is nothing in fundamental science). If we do not appreciate the risky and isolated exploration of new paths, by analyzing contents, and we do not respect the ethics of individual or team research, the passion for knowledge, no institutional engineering will allow us to correctly evaluate our activity, nor to finance scientific projects, which is also necessary; big problems of the ecosystem, for example, would really need it …. But if we do only Big Science, we kill science: No project whose final application is told in advance can never be very innovative. The real novelty, even technical, is always achieved from research that does not imagine it—and often long after or as an indirect finding.

7 Following the Mechanical Rule Versus The Imagination of Configurations of Sense

We are confronted by an important crossroads: extraordinary instruments of interaction and exchange of information can enrich our knowledge and our scientific practices, or they can be employed to normalize us and to make us "follow the rules", to render us all homogeneous. Networks put everyone in an unprecedented position of being able to meet at a distance, of accessing the knowledge of all of humanity in its diversity. The exchange of cultures, ideas, objects, was at the heart of major moments in our history, such as ancient Greece and Renaissance Italy, to mention only two examples. But with this new speed of interaction we could do much better. Or, to the contrary, information networks can be used like "average fields" in physics: With too many neighbors, there are no more singularities, we cannot any more have individuality, we become all grey. We have discussed above the monochromatic images of the world of this type: DNA, the brain, human law, they are all seen as sets of formal rules, instructions and programs, of the same type, like in computers, but enriched in the best case—by some network "noise". An instrument, a computer, an excellent assistive tool used by science to understand the world, in mathematical modeling for example, may also be employed to flatten the world, or, worse, to be identified with the world, a new, common sense of mechanicity. It happens that it also organizes our activities, by computational evaluations of work, bibliometrics being only one example. Rigid medical protocols are yet another terrifying example, as sets of formal instructions. This is the world of pure information, elaborated and transmitted by machines, with no interpretation, nor material body, the flesh at the origin of meaning. Living organisms are always acting, moving, in a protensive gesture: They interpret by this any "friction" on the world, any hit arriving—this is the bodily and material origin of meaning.

The modes of life are profoundly changed by the "everything is information" approaches. We are constantly under the pressure of computing devices that, as many promised since the 1960s, were going to replace man in everything. This was the dream of Classical Artificial intelligence (AI), based on deductive machines, that is on Turing's Logical Computing Machine, as he called it, but well beyond Turing's lucid analysis, (Longo 2018c). Yet, today, the paradigm has changed: The connectionist approach, based on continuous deformations of multi-layer neural nets ("Deep Learning") prevailed, far away from the old "deductive" AI. It is no longer an "imitation" of human intelligence, in Turing's sense (his 1950 paper on the Imitation Game), but a tentative "model" of the morpho-dynamics of the brain—that is of changes of forms, including electric variations and wiring, a model that may implement a highly simplified "learning" process. This is a much more effective approach, much closer to the way animal brains may seem to work. So, not a day passes without hearing: "Attention, do accept no matter which working condition, accept to lose your rights because if not, you will be replaced by machines!"—a voluntary construction of a collective imaginary. This substitution already started since decades, and it continues to happen in offices, and above all, in production chains, since the '70s, in storage centers, today, thanks to old style digital control machines—that is to say in professions which require *reiteration of identical gestures*—the original competence of the digital, Turing-like, machine. Since then, we observe a construction of an imaginary adapted to the subordination to the rules, to mechanical evaluation, to the governance which replaces the government, as observed above. Because also the neural nets machines just elaborate information, as they also are input–output devices. They somehow model the brain, somehow learn, but they are remote shadows of an actual animal brain. This is an always active organ, whose "default" state is a chaotic dynamics. In absence of sensations and perceptions, for example by torturing someone by complete silence, immersed in body-temperature water, in the dark, the brain gets "crazy", its activity goes out of control. The brain is not an input–output machine, but an always super-active organ, *constrained* by a changing context. Its continual activity is canalized by and works only in its preferred ecosystem: The skull of an animal as part of a sensing body in an ecosystem—and in history, as for humans. Its material flesh, the only one we can witness, is essential for this.

Intelligence is not just "elaboration of information", but it is first production of sense, within a biological and historical body in an active friction with the world. Our humanity probably started when we singled-out by interpolation and named non-existing forms of stars' constellations and thus gave "meaning" to the non-sense of the little lights in the sky. Or, when we interpreted phenomena by inventing and naming ancestors and gods, by establishing the law, very differently in different communities. This is not elaboration of information but the "imagination of new configurations of sense", in the stars, in the human community. Our specific, material body and its gestures, in a common history of the communicating community, are essential to this. And these invention are often historical: only the Greek geometers, say, invented the "line with no thickness" (Euclid's definition beta). An abstraction from movement, a trajectory, the invention of the notion of border, which originated the a-logos of the diagonal or the infinity of the Greek π. The extraordinary Nine Chapters of the Chinese geometer Lu Hui, III century, do not imagine this limit construction: The construction of mathematical sense then took a different path. This what I mean by a "historical formation of sense", its reduction to "elaboration of information" is a parody of knowledge construction.

Today, when we speak of constant novelty, certainly, we are surrounded by a million of new fantastic gadgets exchanging bits and bytes of information, but publicity on this matter is often more important than scientific innovation, as hinted above. If you read newspapers from the end of 1990s, you will find similar announcements to those of today's Self-Driving Car … and what has become of the Google Glass in the past two years? It was presented as a revolutionary way to elaborate visual information …. So many promises are continually formulated whose realization ends up as gadgets that decorate our cars, typically. In Pittsburg (PA, USA), a couple of years ago, Uber has launched a big project of driverless taxis, driven by robots that elaborate input information. For the moment, they declare that there will still be an "employee" in the vehicle—as the experiment has been a major failure. Is it possible that such a smart company that earns so much money by just selling information, with no "hardware" (no cars' ownership), is wrong? But no, the primary objective is the increase, in short term, of stock market value of the concerned companies, by the effect of advertising. As for driving automation, we will perhaps channel, by means of electronic tracks, the driving done by robots and humans as well, so that their coexistence on the road can be possible. As long as we are channeled as car drivers, it is great. But the myth of mechanization, by all kinds of algorithms from genetics to evaluation, the necessity of interacting by subordination to machines elaborating digital information—at first theoretical, conceptual and then practical, can have as a goal or as an effect to completely channel our human and social behavior. The too much computational modeling of various forms of "meteorology of social dynamics" seem to go in the same direction: Social information is extracted for the historical context and elaborated by optimization techniques—more than a botched analysis, it seems a method for guidance.

8 Spurious Correlations and Non-Sense in Big Data. Taking a Step Aside

This is the social pressure underlying and justifying the "everything is information" myth. The reduction of human activities and intelligence to "elaboration/transmission of information" by input–output devices continually compares human activities to machines, a fantastic tool when soundly used, but often overestimated in their performances and presented as threatening the human intellect. The worst example is perhaps the new fashion of un-scientific Data Mining on Big Data, as immense reservoirs of information. Big Data and their sound statistical analysis are an opportunity without precedent if they are used to produce hypotheses, to (in-)validate theories and to propose new ones. But, to the contrary, and in a truly viral manner, some think today that we can optimize thought, by reducing it to zero: sufficiently powerful algorithms could in the end "replace scientific knowledge" (sic!). This is supposed to predict all kinds of dynamics and to direct action, without the need of a hypothesis, a theory or knowledge. The bigger a database, yotta of yotta bites, the more, they claim, we can avoid thinking: "machines will discern regularities that the science does not see", but which are sufficient to predict and act (Anderson 2008). "We kill based on Data Mining on metadata" declared the former CIA director, M. Hayden, in a debate (https://www.youtube.com/watch?v=UdQiz0Vavmc).

Luckily, mathematics allows to demonstrate the absurdity of these perspectives. In (Calude and Longo 2016a), by a simple application of classical, not trivial results, we pointed out the "Deluge of Spurious Correlations in Big Data". In short, for any given "correlation between numbers", one can compute a number of elements, let's say *m*, such that *any* set of data (of numbers) having at least *m* elements satisfies the pre-given correlation. It is therefore also the case for a set of numbers, a database, produced by a random process, by throwing of dice or quantum measurements: The correlation will appear there as well and will then be "spurious", because it occurs by chance. In other words, the presence of correlations may only depend on the *size* of the database and do not grant themselves in any case prediction nor taking action. In this way, in order not to think, these authors of algorithms who say they in principle ignore theories, including the ergodic theory, theories of algorithms and of finite combinatorics, that we applied for our result, are confronted with internal limits demonstrated by and within these theories: Spurious correlations inevitably infiltrates very large sets of numbers. This makes any kind of prediction risky when this one is not based on a thought that gives meaning and that enables us to choose what counts, beginning with the choice of what and how to measure and thus to associate it to a number. This thought is necessary to understand and theorize and, if possible, predict. Moreover, the power of scientific knowledge resides also in the fact of revealing the limits of the intended theory; this helps in better understanding, delimiting but also reinforcing the perspective that enables us to do science. Those who pretend to understand everything and to be able to make everything from a single object or concept, such as the DNA in biology or information, algorithms in all sciences … they are certainly wrong. But I insist: the DNA, information, algorithms … are all very important, the *science* around them is essential. The former is a fundamental chemo-physical trace of evolution, continuously employed by cells in order to produce proteins, largely using the Brownian motion of the proteome, which is the first functional random dynamics in biology. The latter, the algorithms, are about to change our lives, potentially for the better, if we keep in mind the limitations of any monomaniacal approach to knowledge (Longo 2005, 2018a).

In this "everything is information" or even "instructions and formal rules" craze, with the peak in the use of Big Data with no meaning, many often make reference to Alan Turing, inventor of the Logical Computing Machine (1936)—the mathematical foundation of computers—but with a dangerously incorrect interpretation of his work, see (Longo 2018c). We need to go beyond a collective imaginary in order to think the "next machine", just as Turing was able to show the falsity of common hypotheses by his result of "non-computability". In fact, between us, this digital machine and its powerful networks are perhaps a little boring, with their always identical iterations: If one initiates over and over a digital simulation of the wildest turbulence, with the same initial data, or if one opens a web page in Japan, they are always the same. But when faced with a non-linear physical dynamics such as a hurricane or a linguistic structure, all this is suddenly meaningless because no hurricane or human being will ever repeat anything identically. When digital information is networked, the sophisticated methods of "interleaving" and "semaphores" render the randomness, proper to the spatial–temporal continuum and of networks, ineffective. This randomness is considered as a "do not care", as the experts—the real ones—say. In the areas of concurrency and network analyses in computer science, experts are able to make them function with computational certainty, according to the rules, eliminating noise. In fact, against computing's common-sense view, which belongs more to molecular biology and to formally regulated governance, computer science today analyses and makes large use of randomness, including in continua, by stochastic calculus, analyses of networks dynamics, in techniques of coding, although in a way different from biology. Networks' noise is thus analyzed and used in a very useful way, in cryptography for example. Yet, this has nothing to do with evolutionary "bricolage" of rare events, the unpredictable changes of the phase space (the space of observable and pertinent parameters), which are proper to biological dynamics and functional to them (Longo 2016a, b). The historical processes in natural and human sciences as well, economics in particular, require a different insight far away from the analysis by and the imposition of the optimal path in pre-given spaces of possibilities, see (Koppl et al. 2015).

It is the job of science to single-out the role of diversity of the human within the human, starting from the biological. Evolutionary diversity and adaptability is the result of the unexpected variant, the "hopeful monster" that is not noise which needs to be removed, like in information theories, or averaged out as in stochastic approaches, but it is that which makes life possible, according to Darwin's great intuition. This may also allow us to better use this fantastic discrete state machine for the elaboration of information, and its networks, and to invent new machines. Human beings will certainly invent yet another machine, if we do not identify this one and its algorithms for elaborating information with the world, if we abandon the necessity to work in the same computational modalities, evaluated, or with our actions predicted, by normalizing techniques without thought, like in the science handled by Big Data and bibliometrics. Making the founding principle explicit, in order to re-think them critically, is at the core of the scientific approach and its ethics. Science is not "elaboration of information", even less the extrapolation from Big Data, but a historical construction of sense of the world, by a continual invention of new concepts and structures. In order to trust science, its ethics requires a critique of one's own principles of knowledge, making a "step aside", understanding different viewpoints. A good scientific practice is a major democratic training by this critical attitude, while contributing to democracy also by knowledge construction. It may only be the result of a difficult dialogue between remote or contrasting perspectives, a collaboration between different knowing subjects, based on ideas that may be unlikely but must be profound, proposed by small groups of thinkers who see things differently. Disagreement and diversity are not noise to be eliminated by the absolute dominance of the majority vote, such as the short term quotations in bibliometrics. They are instead at the core of science, since its invention by the democratic and the scientific debate on the Greek Agora. To do science, we need democracy, and science is an essential component of democracy, the only way to trust both.

References

More in the papers and references in https://www.di.ens.fr/users/longo/download.html

Anderson C (2008) The end of theory: the data deluge makes the scientific method obsolete. *WIRED*. Retrieved from https://www.wired.com/2008/06/pb-theory/

Audétat M (ed) (2015) Sciences et technologies émergentes: Pourquoi tant de promesses? Hermann, Paris

Baker S (2014) Recognizing paradigm instability in theories of carcinogenesis. Br J Med Med Res 4(5):1149–1163

Barbaresco F, Mohammad-Djafari A (eds) (2015) Information. Entropy and Their Geometric Structures, MDPI, Basel and Beijing

Bravi B, Longo G (2015) The unconventionality of nature: biology, from noise to functional randomness. In Calude CS, Dinneen MJ (eds) Unconventional Computation and natural computation, vol 9252. Springer International Publishing, Cham, pp 3–34. https://doi.org/10.1007/978-3-319-21819-9_1

Bizzarri M (2012) The new alchemist. The risks of genetic modification. MIT Press, Boston

Buiatti M, Longo G (2013) Randomness and multilevel interactions in biology. Theory Biosci 132(3):139–158. https://doi.org/10.1007/s12064-013-0179-2

Calude CS, Longo G (2016a) Classical, quantum and biological randomness as relative unpredictability. Nat Comput 15(2):263–278. https://doi.org/10.1007/s11047-015-9533-2

Calude CS, Longo G (2016b) The Deluge of Spurious Correlations in Big Data. Found. of Science, 1–18, March, 2016

Diamanti-Kandarakis E, Bourguignon JP, Giudice LC, Hauser R, Prins GS, Soto AM, Zoeller RT, Gore AC (2009) Endocrine-disrupting chemicals: an Endocrine Society scientific statement. Endocr Rev 30:293–342

Elowitz MB, Levine A, Siggia E, Swain P (2002) Stochastic gene expression in a single cell. Science, 297

Giuliani A (2010) Collective motions and specific effectors: a statistical mechanics perspective on biological regulation. BMC Genomics 11(suppl 1):S2

Koppl R, Kauffman S, Felin T, Longo G (2015) Economics for a creative world. J Inst Econ 11(01):1–31. https://doi.org/10.1017/S1744137414000150

Longo G (2005) On the relevance of negative results. Intellectica (40) (in English: https://www.di.ens.fr/users/longo/files/PhilosophyAndCognition/neg-resCE.pdf)

Longo G (2016a) How future depends on past histories and rare events in systems of life, Foundations of Science, pp 1–32

Longo G (2016b) A review-essay on reductionism: some reasons for reading. In: Chibbaro S, Rondoni L, Vulpiani A (eds) Reductionism, emergence and levels of reality. The importance of being borderline. Urbanomic, London. https://www.urbanomic.com/document/on-the-borderline/, May 8

Longo G (ed) (2017) Lois des dieux, des hommes et de la nature. Spartacus IDH, Paris

Longo G (2018a) Interfaces of incompleteness. In: Minati G, Abram M, Pessa E (eds) Systemics of incompleteness and quasi-systems. Springer, New York, NY

Longo G (2018b) Information and causality: mathematical reflections on cancer biology. Organisms J Biol Sci 2(1):2018
Longo G (2018c) Letter to turing. Theory, Cult Soc, A Posthumanities Special Issue, June 7, 2018 https://doi.org/10.1177/0263276418769733
Longo G (2019) Information at the threshold of interpretation, science as human construction of sense. In Bertolaso M, Sterpetti F (eds) Critical reflection on automated science—will science remain human? Springer, New York
Longo G, Montévil M (2014) Perspectives on organisms: biological time, symmetries and singularities. Springer, Berlin
Longo G, Montévil M, Kauffman S (2012) No entailing laws, but enablement in the evolution of the biosphere. ACM proceedings of the Genetic and Evolutionary Computation Conference, GECCO'12, Philadelphia (PA, USA), July 7–11
Longo G, Seno L (2018) Digital networks, knowledge and "political" biases in their understanding and use. In: Stiegler et al (ed) Philosophy of internet and of the hermeneutic web. MIT Press, in preparation
Montévil M (2020) Entropies and the anthropocene crisis, AI and Society, to appear
Plomin R (2018) BLUEPRINT How DNA makes us who we are, Penguin
Rafols I et al (2014) Big pharma, little science?: a bibliometric perspective on big pharma's R&D decline. Technol Forecast Soc Chang 81:22–38
Seno L (2014) Why the development engine broke down. IEA Nantes. Downloadable from: https://www.acustica.org/publicat.htm#2014
Sikandar H (2019) Genome-editing technologies: concept, pros, and cons of various genome-editing techniques and bioethical concerns for clinical application. Mol Ther: NuclC Acids 16
Sonnenschein C, Soto AM (1999) The society of cells: cancer and control of cell proliferation. Springer
Soto AM, Longo G, Noble D (2016) From the century of the genome to the century of the organism: new theoretical approaches. Prog Biophys Mol Biol 122(1):1–3
Vandenberg LN et al (2012) Hormones and endocrine-disrupting chemicals: low-dose effects and non-monotonic dose responses. Endocr Rev 33:378–455
Weinberg R (2014) Coming full circle—form endless complexity to simplicity and back again. Cell 157, 3/27

Epistemic Inequality and the Grounds of Trust in Scientific Experts

Pierluigi Barrotta and Roberto Gronda

Abstract The goal of this article is to expose the structural complexity of citizens' trust in scientific experts, and to argue for the possibility of a productive cooperation between citizens and scientific experts within a community of inquirers. We firstly distinguish between three different idealized forms of epistemic inequality, with the purpose of shedding light on the distinctive features of the relationship between laypeople and scientific experts. Then, we highlight the multi-layeredness of the layperson-expert trust: though laypeople's trust in science is of an epistemic kind, we maintain that its grounds are rather deontological and institutional. Finally, we show how the radical epistemic inequality between laypeople and scientific experts does not rule out the possibility of public deliberation on public issues, the latter being conceived of as problems in which scientific and socio-ethical–political components are essentially interwoven.

1 Introduction

The study of trust is now a hot topic in philosophy. In recent years, epistemologists, moral and political philosophers, historians of philosophy, philosophers of science have all turned their attention to the analysis of the notion of trust. Other related concepts such as testimony and deference have had a similar fate. The common thread that we see running through all those approaches is the acknowledgment that the different epistemic activities in which humans beings are engaged cannot be accounted for without considering the social dimension in which such activities take place. Epistemic activities are, first and foremost, human activities: as such, they depend on the social environment in which human beings are reared and within which they operate as epistemic agents.

P. Barrotta · R. Gronda (✉)
Department of Civilizations and Forms of Knowledge, University of Pisa, Via Pasquale Paoli 15, 56126 Pisa, Italy
e-mail: roberto.gronda@unipi.it

P. Barrotta
e-mail: pierluigi.barrotta@unipi.it

A. Fabris (ed.), *Trust*, Studies in Applied Philosophy, Epistemology and Rational Ethics 54, https://doi.org/10.1007/978-3-030-44018-3_6

As philosophers of science, we very much appreciate the recent emphasis placed on the role of trust in the process of knowledge acquisition. As Lorraine Code has pointed out, too often in the past, philosophers of science have adopted an impersonal view of scientific knowledge, to the effect that "knowledge in the objective sense is *knowledge without a knower*; it is *knowledge without a knowing subject*" (Code 2017, 725; the quotation is drawn from Popper 1979, 109). To put it simply, the rationale behind this view was that the context of justification is governed by purely logical relations between sets of premises and sets of conclusions. Accordingly, no room was left for the analysis of the different ways in which knowledge is gained.

The collapse of the justification/discovery dichotomy has dramatically altered the landscape of the philosophy of science. Emphasis has progressively shifted from the logical reconstruction of the forms of justification used in natural sciences to the analysis of the epistemic practices that constitute and characterize different research communities. A more empirically-driven approach to scientific knowledge has thus brought to light the importance of trust for the functioning of contemporary science (Hardwig 1991; Whyte and Crease 2010). It is a fact that the division of cognitive labour has now reached a point where no single researcher is capable of providing a satisfactory justification of even the simplest bit of knowledge without relying on others. If scientists did not trust each other, scientific knowledge would be practically impossible.

In more recent times, the debate over scientific knowledge has taken a further step towards a more radical social epistemology. The standard account of the growth of scientific knowledge runs approximately as follows. First of all, scientists discover new relevant truths in their laboratories, relying on the body of knowledge accepted by their communities. Then, they submit their work to the scrutiny of their epistemic peers, asking them for a certification of validity for their results. The bonds of trust that underlie the communities of researchers make scientific knowledge a public good. Finally, scientific knowledge is communicated to citizens and stakeholders, with the purpose of being incorporated in public policies. The higher the trust of citizens and politicians in science, the more effective are likely to be the policies they implement.

Such standard account has been challenged from many sides, and it has been argued that the idea of trust that it advocates is theoretically unsatisfactory. For instance, it is now generally acknowledged that the so-called deficit model of expert communication—which is a corollary of the standard account and, in its turn, underlies the program of the Public Understanding of Science—is based on the over-simplistic assumption that there is only one legitimate form of trust, which goes from citizens to scientists. No genuine epistemic disagreement between citizens and scientists is admissible. According to that model, it is the duty of scientists to communicate scientific knowledge in the easiest and most accessible way; it is the duty of political institutions to promote scientific education; finally, it is the duty of citizens to trust scientists because of the latter's cognitive superiority.

Now, it is indisputable—at least, *we* believe this should not be challenged—that scientists are more competent than citizens in handling scientific knowledge. Nonetheless, concluding from this platitude that citizens cannot play any relevant *cognitive* role in *any* problem-solving process is unwarranted. Think, for instance, of the well-known case-study presented by Epstein in his *Impure Science: AIDS, Activism and the Politics of Knowledge* (Epstein 1996). Or, think of the hostile relationship between Cumbrian farmers and official science after the Chernobyl fallout investigated by Winne (Wynne 1996). In both cases, scientific knowledge was successfully challenged by a bunch of citizens who had a different grasp on the issue at stake. In both cases, citizens succeeded in showing that the way in which scientists dealt with social problems was deeply unsatisfactory in that it completely overlooked some relevant features of the scenario.

The recent sociological and epistemological interest in the interplay between scientific and local knowledge has helped highlight the intrinsic complexity of the relation of trust. Contrary to what was previously believed, trust is not unilateral: citizens should trust scientists, but there are good reasons—good *epistemic* reasons—to hold that scientists, in their turn, should trust citizens. These reciprocal bonds of trust bind together citizens and scientists in a broader community of inquirers (Barrotta 2018).

The goal of this article is to shed further light on the shape that trust takes in this broader context of social inquiry. In particular, we aim to introduce some distinctions that we believe may be useful in clarifying the relation of trust between scientific experts and laypeople. Indeed, the realization of the internal complexity of that relationship has led many scholars to jump at conclusions that we consider theoretically unwarranted and socially harmful. For instance, it has been argued that no real epistemic difference exists between scientific experts and laypeople, or that the difference between them should be traced back to moral and political reasons (Collins and Evans 2017).

We want to make it clear that the idea of the community of inquirers does not lead to sceptical outcomes: we hold fast to the explanatory relevance of the epistemic dimension. However, we believe that the kind of epistemic trust that is at stake in a community of inquirers should be properly qualified. In the first section, we begin by examining the concept of epistemic inequality, with which the notion of epistemic trust is essentially interwoven. Our aim is to highlight the distinctive features of the expert/layperson relationship, and its irreducibility to other forms of epistemic inequality. In the second section, we analyse the structure of laypeople's trust in scientific experts, and we investigate its multi-layered nature. We thus bring to light the internal complexity of the phenomenon of epistemic deference to experts. Finally, in the third section, we deal with the epistemic relationship that holds between laypeople and scientific experts within a community of inquirers. By adopting a pragmatist epistemology, we distinguish between scientific and public problems, and we argue that laypeople are trustworthy epistemic agents as participants in the social inquiries aiming to solve public problems.

2 Epistemic Trust and Epistemic Inequality

We trust when we recognize that we have good reasons to defer to expert testimony,[1] an expert (in its broadest, epistemological sense) being anyone who knows enough in her field to be considered a reliable source of information. Epistemic trust goes, therefore, hand in hand with epistemic inequality.[2] Properly speaking, A knows *p* on the basis of B's testimony if and only if there is some sort of epistemic inequality between them. It would make no sense to say that A knows that there is a cat on the mat, while she is perceiving it, on the basis of B's testimony. Similarly, if A said that she believes that $2 + 2 = 4$ because B said so, we would probably blame her for depending on B's testimony instead of relying on her own cognitive capacity.

Epistemic inequality can take different forms. In its simplest form, it consists of a difference in physical accessibility: B is in a better position than A to get some information. For instance, B is looking out of the window while A is lying on the couch: A knows that there is someone in the courtyard on the basis of B's testimony. A slightly different kind of accessibility is that concerning private or past events: A knows that B has a toothache on the basis of B's testimony, since in normal circumstances the subject is considered authoritative about her mental states. Similarly, A knows that yesterday B bought a new book on the basis of B's testimony, since the subject is considered authoritative about her past actions.

Expertise or competence is another, more complex form of epistemic inequality. Roughly speaking, A knows that *p* on the basis of B's testimony if and only if (1) B is more competent than A about *p* and (2) A believes that B is more competent than she is about *p*. Being more competent is not simply a matter of being in a position to have easier access to a certain content. Being more competent is about having more knowledge, more experience, or more understanding than other people regarding a highly complex set of issues.[3]

Our first thesis is that a clear-cut distinction should be made between these two forms of epistemic inequality. We believe that the rationale behind such distinction is quite clear. In the first three cases, epistemic inequality depends on what Williams called the "purely positional advantage" of one person over another (Williams 2002, 42; see also Gelfert 2014, 219; Gelfert 2011). If A stood up and looked out of the window, she would not need to rely on B's testimony to believe that *p*. Similarly, if

---

[1] In this essay, we adopt a trust-based account of testimony, due to the distinguishing character of the expert-novice relationship. For a general overview of the various forms that such account can take, see (Gelfert 2014, Chap. 8).

[2] Starting from the influential (Hardwig 1985), epistemologists have focused on epistemic dependence. We believe that epistemic inequality is more fundamental, from an explanatory point of view, than epistemic dependence: it is because there is an epistemic inequality between A and B that A is epistemic dependent on B. In any case, since the two notions are so strictly interwoven, they are almost perfectly interchangeable.

[3] We do not want to take a stance here on the issue of whether competence and expertise should be considered as substantial or comparative notions. We have outlined our views on that matter in (Barrotta and Gronda 2019).

A could travel back in time and follow B in the latter's daily activities, she would know *directly* that B bought a book yesterday.

The last case—A knows that B has a toothache on the basis of B's testimony—may look more difficult to accommodate within the "purely positional advantage" framework. The main concern seems to be related to the degree of idealization that it involves. Indeed, if one reframes the issue by saying that, if A were B, then she would have direct knowledge of toothache, it seems we would also have to say that if A (a lay person) were B (an expert), then she would have direct knowledge of *p*, *p* being a bit of knowledge in B's field of expertise. Accordingly, the distinction we have drawn between the two *forms* of epistemic inequality (inequality of access and expert inequality) would be, at most, a difference of degree.

We believe that this is not a foregone conclusion. The idea behind the notion of a purely positional advantage is that the difference between two cognitive performances or states can be accounted for *exclusively* by a different accessibility to certain contents. The implicit assumption is that the cognitive capacities that are involved in these cognitive activities are ubiquitous.[4] Perception (both internal and external) is a prototypical instance of ubiquitous activity; reasoning on very simple issues—like $2 + 2 = 4$—is another. Expert knowledge, on the contrary, is the prototypical instance of knowledge involving cognitive capacities that are not ubiquitous. It is not enough for A to be in B's position in order to know that *p*; in addition to that, she would also have to be endowed with B's cognitive capacities. And this is a further level of idealization. From this perspective, therefore, the distinction holds out.

Now that we have laid out the reasons why epistemic inequality due to positional advantage is to be kept separate from expert inequality, our next step is to argue that a distinction can be made between three forms—three ideal types, actually—of expert inequality. We also argue that those forms differ in the kind of inequality that exists between the two or more subjects involved in the relationship.[5]

The first form of expert inequality—call it specialist-specialist inequality—is that among specialists in the same field. Imagine two historians of the French Revolution, Albert and Georges. Georges tells Albert that he has recently discovered new documents on the death of Marat; he is now sure beyond any reasonable doubt that Charlotte Corday was not Marat's murderer. Georges is a very good historian, and Albert is aware of that. So, Albert decides to trust Georges about the innocence of Charlotte Corday: Albert knows that Charlotte Corday did not assassinate Marat on the basis of George's testimony.

---

[4]The formula "ubiquitous capacities" is modelled after Collins and Evans' notion of ubiquitous expertise: "What we call 'ubiquitous expertises' include all the endlessly indescribable skills it takes to live in a human society; these were once thought of as trivial accomplishments" (Collins and Evans 2007, 16).

[5]A word of clarification may be in order here. As has been said, the three forms of expert inequality that we are going to sketch out are *ideal* types. Accordingly, we by no means hold that it is not possible for a layperson to acquire some kind of interactional expertise that enables her to at least partially assess the truth of an expert testimony. Quite the opposite, the creation of "trading zones" is a distinctive epistemic feature of a sound community of inquirers.

Properly speaking, this is a spurious kind of expert inequality. The point is that if Albert were in George's position, having access to the same documents as those that George read, he would know directly, without depending on George's testimony, that Charlotte Corday did not murder Marat. The same insight underlies the ideal of scientific reproducibility and replicability as one of the main criteria of objectivity: if I had the material resources to repeat the experiment you carried out, I would reach the same conclusions as those you arrived at.

In other terms, this kind of expert inequality is spurious because it can be traced back to—and explained in terms of—the mere positional advantage of one subject over the other. The notion of positional advantage is different from that of *purely* positional advantage in that, unlike the latter, the former does not assume that the cognitive abilities involved in the epistemic activities under consideration *must* be ubiquitous. Nonetheless, once that difference has been acknowledged, there are strong structural similarities between the two forms of positional advantage. In both cases, indeed, the subject who is epistemically dependent can always, at least in principle, withdraw her deference and go check whether her trust was misplaced or not. This means that A knows *p* on the basis of B's testimony for pragmatic rather than for constitutive reasons. There is nothing in A's trust in B that A could not personally and directly verify.

Now, let's consider a different scenario—a case of epistemic inequality between two specialists in two different yet interrelated fields. Anna is a brilliant archaeologist. In her last excavation campaign, she discovered a very interesting artefact—say, a funerary urn—that she has reason to believe dates back to the Hellenistic period. She then decides to send the artefact to a research centre specializing in Carbon-14 dating; the results confirm her initial opinion. Now, Anna knows that her funerary urn dates back to the Hellenistic period on the basis of the testimony of the researchers who did the Carbon-14 dating.

Intuitively, this scenario should strike us as remarkably different from the previous one. In this case, the epistemic inequality between Anna and the researchers who did the Carbon-14 dating cannot be boiled down to a matter of positional advantage. Even though Anna had access to the instruments used to measure the amount of Carbon-14 in her artifact, she could not produce by herself the justificatory evidence that she would need in order to have the knowledge she needs, since she lacks the needed competence. True, Anna could ask for a second opinion, but, again, she then has to trust such second opinion. For this reason, the epistemic inequality between Anna and the Carbon dating specialist is unavoidable: like it or not, Anna has to trust.

But does it mean that Anna's trust in the results of the Carbon-14 dating is to be *blind* (Hardwig 1991, 693)? We think that that would be too strong an assumption. Actually, Anna can advance arguments in support of the belief that *p*—i.e. that the Carbon-14 dating reveals that the remains in the funerary urn she discovered date back to the Hellenistic period. For instance, by relying on her competence in Ancient Greek she knows that the inscription on the urn is not Archaic. So, if the results of the Carbon-14 dating did reveal that remains in the funerary urn dated back to the early period of Greek Civilization, she would be right not to completely defer to it. Maybe, she would hypothesize that the remains were transferred from an Archaic urn to her Hellenistic urn—or something like that. The point is that Anna has good reasons not to accept the belief that *q*, "the funerary urn dates back to the early period of Greek Civilization", though she has no good epistemic reasons to reject *s*, "the Carbon-14 dating reveals that the remains in the funerary urn she discovered date back to the early period of Greek Civilization". Though not a specialist in Carbon-14 dating, she can *actively* and *creatively* participate in the game of giving and asking for reasons, because of her competence in a related field.

Finally, let's consider a case of expert/layperson inequality. John and Jane have to decide about the vaccination of their daughter, Ellen. John and Jane have absolutely no competence in the field of immunology, but since they want to take the best decision for their daughter, they start gathering information. While browsing the web, they come across a blog that recommends vaccine hesitancy: Andrew, the blogger and a physician, states that vaccines may cause autism. John and Jane have to decide whether or not to trust Andrew on *p*—namely, that vaccines may cause autism. Clearly, they cannot directly assess the epistemic status of *p*. Unlike Anna, however, they cannot even advance arguments in support or against *p* on the basis of their competence in a related field, because, *ex hypothesis*, they have none. In this case, the epistemic inequality between John and Jane, on one side, and Andrew, on the other, is so dramatic that John and Jane's trust in *p* cannot be scaffolded by any other epistemic reason.

Clearly, that is not the end of the story. As a matter of fact, John and Jane have a moral duty to avoid epistemic gullibility (John 2011, 509ff.). Now, suppose that, while surfing the internet, they also came across Goldman's influential article *Experts: Which Ones Should You Trust?*, and they were impressed with Goldman's advice to look for *indicators* or *proxies* for scientific competence. It is well known that Goldman's idea is that some indirect criteria may be singled to enable laypersons to "justifiably choose one putative expert as more credible or trustworthy than the other with respect to the question at hand" (Goldman 2001, 92). So, taking Goldman's hints, they decide to assess Andrew's scientific reputation by inquiring into his past track-record, checking his h-index, and so on.

By doing so, John and Jane can actually provide some sort of *indirect* justification for *p*. For instance, should they discover that Andrew is widely discredited by most scientists and physicians, they would have relevant evidence against his credibility as a source of testimony. Consequently, they would have sound epistemic reasons not to trust Andrew: from then on, not only would they not know *p* on the basis of Andrew's testimony; they would no longer believe *p* either.

Thanks to the proxies for scientific competence, therefore, John and Jane too can participate in the game of giving and asking for reasons. It is for this reason that they can be treated as epistemic agents and held responsible for their decisions. What is worth noting, however, is that the way in which they take part in the epistemic game is constitutively different from Anna's one. Indeed, unlike Anna, John and Jane have no kind of epistemic authority over the evidence that they use to confirm or undermine the thesis about the correlation between vaccines and autism. Anna possesses first-hand, direct justification for her beliefs about the inscriptions on her funerary urn. And it is such ability to provide reasons in support of her beliefs that makes it possible for Anna to be an active participant in the epistemic game of giving and asking for reasons.

John and Jane do not have any first-hand justification: of course, they know that Andrew has a low h-index, that he is not respected as a physician, that his institution is not prestigious, and so on and so forth. Nonetheless, they do not have a better grasp on the epistemic reliability of those proxies than they have on the truth of the scientific thesis at hand: for instance, they still have to trust the institutional mechanisms that allocate scientific reputation.

In other words, while it is correct to say that, thanks to the proxies, laypersons can act as epistemic subjects in complex scientific issues, it should also be noted that such epistemic status can only be achieved insofar as, and to the extent that, there are further levels which the responsibility for taking a stance on those issues can be passed off to. In this case, epistemic inequality reaches its highest degree. The expert/layperson inequality can be, therefore, characterized as follows: *trust all the way down*.

3 Layperson-Expert Trust as a Multi-Layered Phenomenon

The previous remarks were intended to pave the way for the idea that the relation of trust between experts and laypeople is a multi-layered phenomenon. In a nutshell, the multi-layeredness of the Layperson-Expert Trust (henceforth LET) refers to the fact that the epistemic trust that laypeople have in scientific experts is the result of a network of relationships of trust that are not, strictly speaking, epistemic. As will be clear in a moment, with this idea we aim to criticize and correct a too narrow account of epistemic trust, which has serious theoretical and practical consequences.

The best way to formulate the idea of the multi-layeredness of LET is in terms of the class of the possible defeaters of A's knowledge that *p* on the basis of B's testimony, A being a layperson and B a scientific expert. In contemporary epistemology, a defeater is something—usually, a proposition or a mental state; we will not take a stance on this issue—that negatively impacts on A's justification for believing that *p* (Moretti and Piazza 2018; see also Costantin and Grundmann 2018). So, for instance, if I am looking at a green tie in a store and then I realize that there is electric lighting in the store, the belief that *q*, "there is electric lighting in the store", is a defeater of the previously held belief that *p*, "this is a *green* tie". Indeed, we know that electric lighting makes blue ties look green.

This simple idea has explanatory relevance. Let's go back to John and Jane and let's take a look at the possible defeaters of their belief that *p*, "there is a correlation between vaccines and autism", which is based on the testimony of Andrew the physician. The first class of defeaters is epistemic. If Andrew had to write in his blog that, due to new scientific evidence, there is proof beyond reasonable doubt that such correlation does not hold, that would count as a defeater of John and Jane's belief that *p*. Consequently, they would not be justified in holding *p*.

Defeaters of the epistemic kind are the most obvious example; they are not the only defeaters, though. Suppose that, following Goldman's advices, John and Jane start to investigate Andrew's scientific reputation: they discover that he got his Ph.D. from a good university, but he did not earn his Ph.D. in immunology; that he has not published any relevant scientific article on the autism/vaccines correlation and that, on top of that, his h-index is very low. In the light of the new evidence, John and Jane are very likely to discard *p*. In any case, they are not justified in believing that *p*.

Now consider a slightly different scenario. Suppose that John and Jane's investigations into Andrew's scientific reputation have found that Andrew is a very good physician. That, coming from a non-affluent family, he graduated in one of the top universities in Europe; he got his Ph.D. from one of the best universities in the USA and was then hired by one of the leading research centres in immunology. He authored a lot of important studies on the effects of vaccines on human health, as a consequence of which he acted as a consultant to several governmental institutions. Since all the institutional proxies unanimously prove Andrew's high scientific reputation, John and Jane are justified in believing that *p*, "there is a correlation between vaccines and autism", on the basis of Andrew's testimony. John and Jane's trust in *p* is well-placed, as far as the institutional proxies are concerned.

However, suppose that John and Jane also know that Andrew is currently under investigation by his institution for mischievous conduct. A scientific journalist got hold of information that Andrew's scientific activity was generously funded by various organizations that have an economical interest in challenging and revising current vaccination programs. This information is a defeater of the belief that *p*. In the light of the new evidence available to them, John and Jane had better not defer to Andrew's testimony. If they deferred, their trust would be misplaced.

We are now in a better position to understand the multi-layeredness of LET. What those two cases reveal, indeed, is that epistemic trust is held accountable to arguments that are not, strictly speaking, epistemic. John and Jane's trust is clearly epistemic, in that it concerns the content of *p*, the alleged correlation between autism and vaccines. And yet the grounds for their trust are not (and cannot be) epistemic, but rather deontological and institutional.[6]

This point can be clarified by referring to the so-called standard account of trust as a three-place relationship. Though a few attempts have been recently made to question the validity of this model,[7] it is usually held that epistemic trust should be conceived of as a triadic relationship between the truster, the trustee, and a proposition, which is the object of trust. Now, in a LET scenario, it is *in principle* impossible for the truster to assess the epistemic authority of the trustee without relying on proxies. The trustee is acknowledged as epistemically trustworthy on the *sole* basis of her deontological value and her scientific reputation.[8] Clearly, we are willing to admit that deontological

[6]A similar approach to trust, which recommends a shift from the epistemic to the non-epistemic (moral) level, has been advanced in (Faulkner 2014).

[7]See, for instance (Domenicucci and Holton 2017).

[8]We use the term "deontological" to refer to the rules that a scientist is expected and asked to follow in order to be a *good* scientist. Merton's CUDOS norms (communism, universalism, disinterestedness,

value and scientific reputation are reliable proxies for scientific competence. But this ultimately depends on the trustful assumption that the institutional mechanisms that govern and control the distribution of scientific reputation are epistemically reliable. It is not difficult to imagine a situation in which such trustful assumption would be misplaced. Think, for instance, of the Institute of Genetics under Lysenko and its mechanisms of attribution of scientific reputation. In that case, laypeople would not be justified in believing that *p* on the basis of the testimony of one of the scientists working in that Institute, no matter whether *p* were false or not.

Generalizing on those previous remarks, we do not simply hold that, in an expert/layperson scenario, deontological and institutional considerations are capable of influencing the justification that a subject has for believing that *p*; more radically, we also argue that deontological and institutional considerations are the *only* factors that influence the epistemic justification a layperson can advance for *p*. The radical epistemic inequality that exists between scientific experts and laypeople makes it impossible for the latter to provide direct epistemic reasons in support of *p*. What they can do is to provide sound arguments in support or against the trustworthiness of scientific experts, by trusting in the reliability of proxies for scientific competence that are made available to the public.

This thesis has remarkable consequences. Here we will only mention a couple of them. On a theoretical level, our account of LET is a strong version of social externalism. We hold that laypeople *know* on the basis of scientists' testimony, since they defer to the latter as to the justification for scientific beliefs. Clearly, social externalism goes hand in hand with anti-intellectualism: with this label, we mean the fact that LET cannot be boiled down to its epistemic component.

The acknowledgment of the multi-layeredness of LET can help shed some light on the practical and theoretical inadequacy of the traditional strategies for dealing with the (controversial) phenomenon of growing distrust in science. The standard solution to this problem has been to enhance the scientific literacy of the public, in the belief that more educated citizens could be more confident in the validity of scientific results. However, empirical research has shown that scientists' trustworthiness has little to do with their scientific competence: scientific experts are not distrusted because citizens believe they are not competent in their field of expertise; rather the opposite, they are distrusted because citizens believe they do not care about the things that matter the most to them.[9].

Accordingly, any attempt to deal with the public's distrust in science in a purely epistemic manner is misdirected. All those programs—think, for instance, of the Public Understanding of Science—are overly intellectualistic, in that they do not adequately appreciate the fact that the reasons of the public's distrust in science and/or in scientists are mainly not epistemic, and therefore they end up recommending practical measures that are bound to be deeply ineffective. Our proposal is to radically shift the attention from the epistemic level to the deontological and institutional

organized skepticism) are a very good example of scientific deontology. For an analysis of the notion of deontology, please see (Fabris 2018, 14ff.).

[9]See, for instance https://tomstafford.staff.shef.ac.uk/?p=447. See also (Sayer 2011).

factors that scientists' trustworthiness ultimately depends upon. We suggest that the citizens' distrust in science and in scientists cannot be dealt with by reducing the epistemic gap between citizens and scientific experts. It has to be dealt with by acknowledging the radical epistemic inequality between scientific experts and laypeople, and then finding a way to enhance the latter's trust in the deontological value of the former, as well as in the reliability of the institutional mechanisms for the allocation of scientific reputation.

4 Citizens and Experts at Play

One of the consequences of the radical epistemic inequality between laypeople and scientific experts that we have stressed thus far is that it is not possible for the former to participate in the production of scientific knowledge. If citizens can be said to know scientific truths only vicariously, it seems that they cannot play the role of active epistemic agents in a scientific enterprise. One may, therefore, be left wondering how our account of LET can be reconciled to the idea of bonds of epistemic trust binding together citizens and scientists in a broader community of inquirers. In this final section we would like to address this question.

The idea of a community of inquirers is drawn from (Barrotta 2018). In a nutshell, it amounts to assuming that scientists and laypeople can cooperate in a common cognitive enterprise, whose objective is truth (Barrotta 2018, 161). Within the pragmatist framework that we also embrace, true beliefs are those that have proven effective in solving a problem and do not create—to use Peirce's words—the irritation of doubt.

The pragmatist emphasis on the conceptual couple doubt/inquiry is pivotal: pragmatists believe that any inquiry is an attempt to solve a *specific* problem, and that the distinctive features of a specific problematic situation also determine the distinctive traits of the inquiry undertaken to get rid of the irritation that the doubt provokes. Having said this, the problem that we have to deal with is, therefore, that of clarifying the nature of the problem that calls up the common cognitive enterprise that encompasses laypeople and scientific experts.

Before attempting this task, however, it is worth spending a few words on a line of reasoning which is quite influential today, in particular among sociologists of expertise, but which we do not believe is the best account of what a community of inquirers is. As has been hinted at above, one possible way of clarifying the idea of a community of inquirers encompassing laypeople and scientific experts is through the notion of local knowledge. In his *May the Sheep Safely Graze? A Reflexive View of the Expert-Lay Knowledge Divide*, Brian Wynne convincingly argued that there may well be groups of citizens that possess a non-scientific knowledge, which is nonetheless relevant to solving a certain social problem. In the case-study presented in that article, Wynne describes the different stages of the process of social inquiry aimed at solving the problem of radioactive contamination caused by the 1986 Chernobyl Fallout. His goal is to show that Cumbrian sheep farmers had reliable knowledge of the local environmental conditions and sheep behavior, which scientific experts deliberately ignored, thus preventing themselves from reaching a satisfactory solution to the social problem they had been asked to tackle (Wynne 1996).

We want to make it clear that we are willing neither to criticize the idea of local knowledge, nor to argue that cooperation between scientific experts and groups of citizens possessing local knowledge cannot give rise to a genuine community of inquirers. This is so far from being the case that one of the authors of this paper drew up his own account of local knowledge (Barrotta 2018; Barrotta and Montuschi 2018). Our point is, rather, that in cases like the one presented by Wynne the bonds of trust binding together laypeople and scientific experts in a community of inquirers come too easily: it is evident and unquestionable that scientific experts have to trust laypeople if the latter have reliable knowledge of some relevant aspects of the problematic situations. But it has to be noted that, though informal and mostly tacit, local knowledge is a form of competence, which is not ubiquitously distributed in society.

Accordingly, we are perfectly fine with the idea of local knowledge, as well as with the prescription to integrate local and scientific knowledge into a more comprehensive course of investigation. We just want to highlight that not every social problem is of this kind. Take, for instance, the problem of deciding whether or not to implement universal vaccination policies. In cases like this, there certainly are groups of laypeople who possess local knowledge—for instance, family doctors who know their patients' behaviour. However, it seems reasonable to include in the community of inquirers that has to deal with that problem all the citizens who will be concerned by its final decision as well.

Public problems of this kind are particularly relevant, because they are the most challenging for those theoretical approaches that aim to avoid the extremes of epistemic relativism—laypeople and scientific experts are on a par—and scientism. In the light of what we have argued thus far, we cannot endorse the view that laypeople and scientific experts are equally authoritative on scientific issues. Because of the radical epistemic inequality that exists between them, it would make no sense to hold that scientific experts should trust laypeople on matters concerning their fields of competence. At the same time, however, it is important to note that public problems do not boil down to their scientific components: the problem of deciding whether or not to implement universal vaccination policies is structurally and essentially different from that of ascertaining whether or not vaccines cause autism. While the latter is solely a matter of specialists, the former concerns the whole public.[10]

Our notion of a broad community of inquirers is rooted in that intuition. The irreducibility of public problems to their scientific components implies that the relevant information that is necessary to deal with them rationally is more complex and more multidimensional than the set of scientific truths delivered by scientific experts *and* the bits of local knowledge possessed by specific groups of laypeople. Clearly, public problems thus understood present a singular entanglement of facts and values: the

[10]For a more in-depth discussion of the distinction between scientific and public problems, see (Gronda 2018).

desires and values of the public are as relevant as factual knowledge to the satisfactory solution of the public problem that called out the social inquiry. The presence of values notwithstanding, the whole enterprise is distinctively cognitive, in that it revolves around the discovery of the means that are best suited to lead to the desired end—that is, a plan of action that could reconstruct the problematic situation, thus getting rid of the irritation that caused the inquiry.

In this scenario, the reciprocity of epistemic trust is brought to the fore. Since all the information relevant to the case under investigation has to be taken into account in order for a social inquiry to be epistemically sound, scientific experts have to recognize laypeople as legitimate and trustworthy epistemic agents insofar as they participate in the problem-solving activity and contribute to shaping the best possible solution. To say that scientific experts have to trust laypeople amounts, therefore, to acknowledging the latter's epistemic authority on the public issues at hand, without implying that they have any epistemic authority on the scientific aspects connected to those issues.

The higher the complexity of the problem to be dealt with, the higher the internal articulation of the community of inquirers. And the higher the internal articulation of the community of inquirers, the tighter the bonds of epistemic trust that bind together the players involved in a common activity of social inquiry.

5 Conclusion

The goal of this article was to highlight the structural complexity of citizens' trust in scientific experts, with the purpose of paving the way for the possibility of a broad community of inquirers encompassing both of them. We have argued that a community of inquirers can be viewed as an epistemic agent engaged in a cognitive enterprise, whose objective is solving the problem that called up the social inquiry. We have also argued that the bonds of trust that bind together the different members of the community are deontological and institutional, rather than properly epistemic.

Clearly, this does not imply that those are the only bonds of trust that hold the members of a community together. Public deliberation is a complex and articulated matter, which depends—among other things—on the legislative regime where that deliberation takes place, as well as on the specific features of the subjects involved in it. All these aspects deserve to be investigated thoroughly. In this sense, our approach welcomes any empirical investigation that could provide further details on the relationship between laypeople and scientific experts.

Our article does not purport, therefore, to formulate a general theory of scientific expertise rooted in the notion of epistemic trust. It aims, more modestly, to outline a normative framework that meets certain requirements—namely, to avoid the twofold threat of scientism and epistocracy—and, in doing so, it expects to help clarify those conditions that are most likely to result in a rational public deliberation.

References

Barrotta P (2018) Scientists, democracy and society: a community of inquirers. Springer, Berlin-New York

Barrotta P, Gronda R (2019) Scientific experts and citizens' trust: where the third wave of social studies of science goes wrong. Teoria 39(1):9–27. https://doi.org/10.4454/teoria.v39i1.54

Barrotta P, Montuschi E (2018) Expertise, relevance and types of knowledge. Social Epistemology 32(6):387–396. https://doi.org/10.1080/02691728.2018.1546345

Code L (2017) Rhetoric and social epistemology. In: MacDonald MJ (ed) The Oxford handbook of rhetorical studies. Oxford University Press, Oxford, pp 721–732

Collins H, Evans R (2007) Rethinking expertise. Chicago University Press, Chicago

Collins H, Evans R (2017) Why democracies need science. Polity Press, Cambridge

Costantin J, Grundmann T (2018) Epistemic authority: preemption through sensitive defeat. Synthese. https://doi.org/10.1007/s11229-018-01923-x

Domenicucci J, Holton R (2017) Trust as a two-place relation. In: Faulkner P, Simpson T (eds) The philosophy of trust. Oxford University Press, Oxford, pp 149–160

Epstein S (1996) Impure science: AIDS, Activism and the politics of knowledge. University of California Press, Berkeley-Los Angeles-London

Fabris A (2018) Ethics of information and communication technologies. Springer International Publishing, Cham

Faulkner P (2014) A virtue theory of testimony. Proceedings of the Aristotelian society 114:189–211. https://doi.org/10.1111/j.1467-9264.2014.00370.x

Gelfert A (2011) Expertise, argumentation, and the end of inquiry. Argumentation 25:297–312. https://doi.org/10.1007/s10503-011-9218-7

Gelfert A (2014) A critical introduction to testimony. Bloomsbury, London-New York

Goldman A (2001) Experts: which ones should you trust? Philos Phenomenol Res 63(1):85–110. https://doi.org/10.2307/3071090

Gronda R (2018) Rethinking the notion of public. a pragmatist account. In: Barrotta P, Scarafile G (eds) Science and democracy: controversies and conflicts, John Benjamins Publishing Company, Amsterdam pp 53–70. https://doi.org/10.1075/cvs.13.05gro

Hardwig J (1985) Epistemic dependence. J Philos 82(7):335–349

Hardwig J (1991) The role of trust in knowledge. J Philos 88(12):693–708

John S (2011) Expert testimony and epistemological free-riding: the MMR controversy. Philos Q 61(244):496–517. https://doi.org/10.1111/j.1467-9213.2010.687.x

Moretti L, Piazza T (2018) Defeaters in current epistemology. Synthese 195(7):2845 2854 https://doi.org/10.1007/s11229-017-1551-4

Popper R (1979) Objective knowledge: an evolutionary approach. Clarendon Press, Oxford

Sayer A (2011) Why things matter to people. Social science, values and ethical life. Cambridge University Press, Cambridge

Whyte KP, Crease RP (2010) Trust, expertise, and the philosophy of science. Synthese 177(3):411–425. https://doi.org/10.1007/s11229-010-9786-3

Williams B (2002) Truth and truthfulness. Princeton University Press, Princeton

Wynne B (1996) May the sheep safely graze? A reflexive view of the expert-lay knowledge divide. In: Lash S, Szerszynski B, Wynne B (eds) Risk, environment & modernity. Towards a new ecology. SAGE Publications, London, pp 44–83

On Trust in Mathematics: Some Case Studies

Luca Bellotti and Enrico Moriconi

A good proof is one that makes us wiser
Yu. I. Manin

Abstract In this paper we consider the role of trust in relation to mathematical proof. After some basic remarks on the main philosophical issues involved, we give material for five case studies, dealing in particular with a few recent disputes on controversial proofs (or purported proofs) of important conjectures. We try to assess some criteria of evaluation in order to obtain a possible philosophical lesson.

Keywords Trust · Mathematics · Proof · Proof assistant

1 Introduction

Our aim in this paper is to take seriously the role of trust as far as the theory and practice of mathematical proof is concerned. In the second section we shall discuss some of the many philosophical issues at stake, showing how pervasive and philosophically problematic this role is. In the third section we give some material for five case studies: we take some recent disputes on controversial proofs which have been proposed for some famous conjectures. Finally, in the fourth section, we come back to the problems considered before, trying to see whether some possible, sensible criteria of evaluation of the disputes emerge.

L. Bellotti (✉) · E. Moriconi
Department of Civilizations and Forms of Knowledge, University of Pisa,
Via Paoli 15, 56126 Pisa, Italy
e-mail: luca.bellotti@unipi.it

E. Moriconi
e-mail: enrico.moriconi@unipi.it

A. Fabris (ed.), *Trust*, Studies in Applied Philosophy, Epistemology
and Rational Ethics 54, https://doi.org/10.1007/978-3-030-44018-3_7

2 Philosophical Issues

The seminal work by Lakatos, *Proofs and Refutations* (1976), displays many themes which are central for our investigation. Through a penetrating criticism of the concept of formal proof—i.e. of an argument executed according to the rules of a precisely specified mechanism—he aims to give some real sense to the claim that regimenting proofs in order to clarify their assumptions and the procedural rules involved—the very process which formalization idealizes—is just *one* phase in the complex process that leads to the growth of mathematical knowledge. In opposition to the *traditional perspectives*, the designation he uses to refer to both Euclidean and Hilbertian kinds of axiomatics, Lakatos rejected the idea that mathematics boils down to the search for absolute proof. Bringing to the fore how important is the activity of trial and error, he proposed as more true to the facts the framework in which the starting point is got by adapting a somehow devised proof-sketch to new problems waiting for a solution. A main point, in this framework, is given by the way in which the evolution of the initial proof is depicted as something which results from interactions with various kinds of counterexamples which immediately start to crop up, leading to arguments over the meaning of terms involved in the definitions as they are put forward, so that various characterizations of the notions of polyhedron, polygon, edge, area, vertex, ... are provided. On this basis he intends to assimilate mathematical theories to general scientific theories, turning them into *quasi-empirical theories*. In the *Introduction* to *Proofs and Refutations* Lakatos argues that he will challenge mathematical formalism elaborating

> the point that informal, quasi-empirical, mathematics does not grow through a monotonous increase in the number of indubitably established theorems but through the incessant improvement of guesses by speculation and criticism, by the logic of proofs and refutations (ibid., 5).

Lakatos concludes emphasizing that sealing a proof with the motto *QED* is unsuitable, since more often than not the final formulation is considerably different from the one which was the starting point. This fact converges to the belief that proofs, and consequently theorems, are becoming a sort of "social entities". Theorems lack a true proof, but circulate endowed with proof-sketches, with hints of argumentative lines, suggestions, ... which constitute a sort of *sociological* validation of proofs which are rated adequate to provide a *sufficient* guarantee of the soundness of the result. At least, as it is said, *until proven otherwise*.[1]

There is a second theme which is relevant for us, even though in a sense probably different from Lakatos's intention. Again in the *Introduction* to *Proofs and Refutations*, he says that once mathematics has been identified with formalized mathematics, there are just two kinds of *discoveries* one can make:

> *First*, one can discover the solution to problems which a suitably programmed Turing machine could solve in a finite time (such as: is a certain alleged proof a proof or not?). No mathematician is interested in following out the dreary mechanical "method" prescribed by such

[1]For a thorough discussion of these points one can see Bundy et al. (2005).

> decision procedures. *Secondly*, one can discover the solutions to problems (such as: is a certain formula in a non-decidable theory a theorem or not?), where one can be guided only by the "method" of "unregimented insight and good fortune" (ibid., 6),

adding that this harsh alternative between "the rationalism of a machine" and "the irrationalism of blind guessing" is not the necessary output of the development of mathematics. As regards the latter point, one can say that, notwistanding the wonderful exploration of how informal mathematics evolves that he displayed in *Proofs and Refutations*, the work to get the searched new kind of "logic", or "method", going beyond the peculiarities of the specific treatment of Euler's conjecture, is still fundamentally lacking.

More difficult is to provide a balanced assessment concerning the former point. On the one hand, no doubts, what Lakatos rates to be a boring and uninteresting job represents instead a central feature of actual mathematical practice. However, a widespread belief is that formalisms in general are simply not a practical tool. Many people concede that formalisms are essential to mathematics, but at the same time they stress that their being essential boils down to what could be called *in-principle formalizability*. So people suspect that 'if the in-principle part of in-principle formalizability is so essential, perhaps there is something wrong with the notion of formalizability in the first place' (Alama and Kahle 2013, 149).[2] The last comment usually matches a sort of intolerance towards formal logic, considered just as the provider of rigid scaffoldings. This opinion, however, though popular, seems also debatable: somebody, in fact, could suggest that the dialectic *informal-formal* proofs could only take advantage from the repercussions that the "in-principle" results could have on the "practice" of getting and writing proofs. To take a trivial example: if somebody says that she crossed all the Königsberg bridges, then we know that she crossed at least one bridge twice. More generally, it is no doubt widespread the opinion according to which the *formal* proofs provided by means of logical tools are mostly useful to enable the implementation of *proofs about proofs* (like in the celebrated Gödel's Incompleteness Theorems). This restrictive perspective is in part misleading and we would suggest that it could be better to frame mathematical practice not on the bare distinction "informal-formal", but keeping into account a third, intermediate level: (informal) axiomatic theories. The step from this level to the formal one reflects of course the remarkable change occurred between the end of 19th century and the thirties of the 20th century, when formalization came to the fore. Frege's and other people's discovery of the possibility to formalize mathematical knowledge produced in fact a new theoretical subject: *formal theories*. Here, "formal" extends its meaning, since

- On the one hand, it means that a given part of mathematical knowledge undergoes a process of rebuilding which aims at making explicit its logical structure so that the justification procedure of informal mathematical proofs can be fully understood.

[2] See also Alama's *Formal proofs and refutations*, unpublished Ph.D. Thesis, Stanford University, 2009.

- On the other hand, "formal" means that we give up using natural languages and adopt conventional artificial ones, without any particular meaning assigned to (strings of) the signs of the language.

Far from conflating axiomatization and formalization, in this way the two operations are kept distinct, stressing that the latter notion refers to the second meaning of "formal"; and this, of course, even though axiomatization is a *necessary* condition of formalization, and even though any axiomatization exploits a (previous) work of formalization (referring in this case to the first sense of "formal").

The construction of a formal theory is not a simple work of *decoration*. On the contrary, it produces new questions, new knowledge, and new standpoints: in a word, it produces a growth in mathematical knowledge. The third layer marks the point where the system, beyond being the framework where investigation is developed, becomes an "object" of study in itself. Of course, this was the theoretical condition which made it possible to get the incompleteness results, and also the development (by Tarski) of a formal theory of the way in which *truth* can be attributed to a sentence of a formal mathematical theory. Moreover, it opens the way to new questions: for instance, one could wonder whether when a *formal* proof of a mathematical sentence is available, we gain some knowledge which goes beyond knowing that the sentence is provable (in a *pre-formal* theory, for instance). We can furthermore argue about, and elaborate on, both the *proof* of a sentence—by providing a *normal* proof, or getting a speed-up, or extracting pieces of computational data from the proof of a sentence in first-order Peano Arithmetic—and its *provability*.

2.1 Changes in the Concept of Proof

Leaving aside this kind of speculations, more pertinent to our question is perhaps to focus on the evolution occurred in the concept of proof in contemporary mathematics. Here "contemporary" means something like "in the last sixty-seventy years", and "evolution" hints, in the first place, at the gradually more and more important role played by computers.[3] This involvement in not limited just to supporting reasoning or calculation steps in proofs, but includes the setting up of programs able to check mathematical results. It goes without saying that in this way a deep tension enters mathematical activity: in order to be submitted to computer-aided checking, proofs are to be written in a completely formal language, something close to the Fregean completely analyzed language. On the other hand, as foreseen and stressed

[3]The procedure that Appel, Haken and Koch offered and that was accepted as a proof of the Four Color Theorem dates back to 1976; the first involvement of computers in mathematical research, however, was in the 1950s. On the whole subject one can see Hales et al. (2008) and MacKenzie (2001).

by Lakatos, most actual proofs that are presented by the mathematical community are very far from that ideal. They contain gaps, they make (sometimes in an essential way) reference to other results which on their turn sail in the same sea, and so on and so forth. Not to mention the fact that a large amount of theorems—which are collectively referred to by the name *mathematical folklore*—circulate without any proof appearing in print.

> Consequently, it becomes increasingly complex to correct mistakes in the literature, since new results are based on other results that are potentially flawed and potentially insufficiently checked by either reviewers or authors of the new article (Frans and Kosolosky 2014, 353).

As a result, proofs are often presented in a way that makes their surveyability and verifiability far from trivial. From a different perspective, this same effect is produced also by computer-aided or computer-assisted proofs (in which the computer is essential in generating the proof) and by computer-based proofs (in which the computer is needed in the proof itself, as in the previously mentioned Four Color Theorem), and by extremely long proofs (as the approximately 10,000-page "proof" of the Classification of finite simple groups). The extreme increase in specialization and complexity of proofs is a third source which contributes to making people aware of the deep difficulties of the view which characterizes the epistemic uniqueness of mathematics in terms of possessing what is now considered an *old-fashioned* property: certainty.[4]

Thus, there are proofs which lack one or more of the important ingredients of that certainty: as, for instance, the fact that a proof must be surveyable, that it can be given in an *actual written instance*, that it can produce knowledge, which in turn requires that it should be comprehensible.

> No doubt there can be "theories" with hundreds of primitive operations and hundreds of axioms about them and programs can be written to correctly derive consequences of these axioms. But these will (mostly) not be proofs since the claims derived will be incomprehensible, and hence unknowable, to agents like us. Such purported proofs might not be long, but they would be in another way too complex to be comprehensible. We might want to rule them out and for the same reason some long "proofs" might need to be ruled out (Coleman 2009, 32).

A good proof should also be *verifiable* step by step. Obviously, "verifiability", or "surveyability" for that matter, are questions of degree: proofs have these properties to some degree for a specific community of mathematicians. They, however, are independent properties. The Appel Haken proof of the Four Color Theorem is (in a sense) not humanly verifiable, but it is not unsurveyable. This depends on the fact that from the proof it is possible to extract a perspicuous representation of the proof *idea*. And this is enough. As Coleman rightly concludes (ibid.): 'The product of a survey is the map, not the territory'.

[4] As Ruelle (2000, 254) comments in an incisive way, 'human mathematics is a sort of dance around an unwritten formal text, which if written would be unreadable'.

2.2 The Role of Referees

A collateral, but absolutely not unimportant, consequence of the above mentioned changes concerns the role of referees. Checking a submitted paper has become a very arduous work sometimes, and it is increasingly difficult to find people available and competent to make it. To encourage scholars to be involved in the refereeing practice, many people made the proposal to separate the fundamental and laborious work of checking proofs from the "scientific" task to judge, on the basis of the reports produced by the referees, on the originality and publication worthiness of a paper. It goes without saying that the latter job should continue to be done anonymously. Instead, as regards the former kind of work, it is very difficult to understand why it should be bound to be blind. Checking the steps of long and/or difficult proofs, assessing the correctness of the general framework, just to mention two features of the work, are very important tasks and not rarely they contribute substantially to get a good paper or book.[5] This is a sensible proposal, since in this way people engaged in the refereeing process could be granted credits for the work done. It is hoped that the possibility to earn academic points would encourage mathematicians to devote part of their time to the task of checking proofs. This is a good proposal, even though some details are still in need of settlement. For instance, one could wonder why, once the two parts of the refereeing work have been separated, an author should modify his or her paper according to the recommendations of the *checking* referee, without having any guarantee that the *scientific* referee will rate the paper worth to be published.

2.3 Trust

One of the conclusions to be drawn is that a lot of texts and information circulate without having been verified; hence they need to be *trusted*. Since trust, being at least partially blind, involves risk, it has been a natural tenet of epistemology that 'knowledge rests on evidence, not trust' (Hardwig 1991, 693).[6] Against this unyielding traditional position, a further element emerged in the last decades, an element not taken into account before: that is, the fact that it is very near its extinction the species of papers which are the outcomes of a *single* author. We already pointed to the debts often owed by an author to many other preceding ones, but a new, greatly increasing form of dependence is due to the fact that nowadays research is increasingly a matter of larger and larger teams. Considering what we said before, this fact is certainly not surprising: the process of gathering data, making experiments, connecting different specialized knowledge and so on and so forth is a work that cannot be accomplished

[5] It is very usual to find in published papers a footnote in which *anonymous* referees are acknowledged for their important contribute to get an improved version of the work.

[6] Hardwig (1991) is now a "classic" paper on the subject; to a related topic he had already devoted the paper (1985).

by a single person. Of course, many times people co-working in a team shared the same formation and therefore they know each other very well. More often than not, however, participants in the team will have different origin and education, hence possibly different criteria of acceptability of gathered information, and consequently also different positions about whom they trust. The resulting picture is that the rationality of many of our beliefs and scientific acquisitions depends on what others do and know, with the consequence that it is not completely within our own individual control.

Accordingly, Hugo Mercier and Dan Sperber in various works[7] elaborate on this feature and emphasize that trust needs to be buttressed by a variety of mechanisms of epistemic vigilance. *Reason* is one of these mechanisms. If humans could systematically trust one another's testimony and advice, they wouldn't need reasons to be convinced. Reasons take over when trust is insufficient to get a message across.

Not surprisingly, the philosophical and epistemological debates on trust have opened a space also for logical investigations devoted to provide a formal characterization of the trusting relation. Literature on this subject is growing continuously, developing calculuses and the pertinent semantics both in the area of proof-theoretic approaches and in the vein of model-theoretic, Kripke- and dynamic semantics of modal logics.[8] It would be out of place to try to detail this evolving work here. Worth to be mentioned is however the debate about the difference in identifying trust as a first-order relation between agents (agent A trusts agent B) or as a second-order relation (relation X between agent A and agent B is trustworthy) and to determine on that basis the relevant formal structure. We conclude by mentioning only the description of *untrustworthy* relations, i.e. relations qualified by a negative assessment of trust. Defining the *negation* of the relation $trust(A, B)$ simply as $\neg trust(A, B)$ would entail giving up the possibility to separate distrust in *any* message from a given agent and distrust in *some* message. The study of this situation on the basis of a model of trust as a second-order property of relations allows one to disentangle the various threads, so that it is possible to provide a more fine-grained conceptual analysis of distrust and mistrust. Following Primiero and Kosolosky (2013), one may say that

- an information transmission that generates uncertainty in the receiver concerning the truthfulness of its content is *untrustworthy*;
- an information transmission that is deemed to convey unintentionally false information is *mistrustful*;
- an information transmission that is deemed to convey intentionally false information is *distrustful*.

A receiver that assesses an information transmission as mistrustful or distrustful operates on its content accordingly.

---

[7]We just recall Mercier and Sperber (2019).

[8]As regards the former area, we mention only Primiero and Kosolosky (2013) and Primiero and Taddeo (2012).

3 Material for Five Case Studies

3.1 The Hales Case

The Kepler conjecture says that no packing of congruent balls in Euclidean three-space has density greater than that of the face-centered cubic packing (the one used by greengrocers). Telling the story of his own proof of the conjecture, Thomas Hales says something of the greatest interest with respect to two prominent issues discussed above, namely the use of computers in mathematics and the role of referees. Hales explains that the anomalous delay in publishing his (informal) proof of this conjecture, obtained in 1998 and published only in 2006, was due to the fact that the referees had to verify in fact a complex computer proof: a proof such that it was not really possible for a human being to check every step. Thus, at the end of the reviewing process, the referees could only reach a strong degree of conviction that Hales's approach was correct, and 'the proof was published without complete certification from the referees' (Hales et al. 2017, 1). It was in 2003 that Hales, thus, proposed a long-term project to give a *formal* proof of the Kepler conjecture: the Flyspeck project. The idea was to formalize down to the least detail both the informal, classical sections of the proof and its coded computational parts. Hales et al. (2017) is precisely the final official account of the project (which was completed in 2014), which has probably the largest number of lines of code ever in a verification project. It was carried out by means of two different proof assistants, HOL Light and Isabelle, with a further verification of the main statement (which reduces the conjecture to three subclaims needing long computations to be checked—the third requires 5K CPU hours!) in HOL Zero, by employing many computers in a 'cloud computation'. Nevertheless, the authors emphasize that 'a formal proof in the HOL Light system is *more reliable by orders of magnitude than proofs published through the traditional process of peer review*' (ibid., 3, emphasis added). In a final section of the paper ('Auditing a distributed formal proof'), they add that 'a proof assistant largely cuts the mathematical referees out of the verification process' (ibid., 24). The human eye is still needed, of course, but only at some spots a specialized mathematical expertise plays a role, while 'the rest of the audit of a formal proof can be performed by any trained user of the HOL Light and Isabelle proof assistants' (ibid.). The auditor cannot simply rely on the improbability of an innocent error, but must 'assume malicious intent'. Especially with this particular formal proof, which was not obtained in a single session of a single proof assistant, very careful auditing was mandatory.

Some (maybe all) reasonable necessary conditions in order to be confident in the results of computations in proofs were listed already by Detlefsen and Luker (1980) in an old discussion of the Appel–Haken proof (1976) of the Four color theorem: '(a) that the underlying algorithm to be used is mathematically sound; (b) that the particular program to be used is a correct implementation of this algorithm; (c) that the computing agent correctly executes the program; (d) that the reported result was actually obtained' (Detlefsen and Luker 1980, 808).

3.2 *The Knight Case*

The Vaught conjecture says that the number of countable models in an axiomatizable class of structures is either countable or continuum. The consensus is that the conjecture is still open, but in 2002 Robin Knight proposed a counterexample. His drafts were thoroughly discussed in a seminar at Berkeley, and the conclusion was that 'the draft of January 2003 does not contain a correct disproof of Vaught's conjecture and requires more than minor emendations to produce a complete proof' (Scanlon 2010; we were unable to find more recent reliable evaluations of Knight's attempt). The trouble was not a specific mistake, rather a lack of apparently necessary technical details. Significantly, Scanlon adds: 'It is possible that revisions he has posted since then are sufficient, though in view of how much time it would require to enter into the details of the argument, I am not willing to work through the later papers *until the basic architecture of the proof is certified by some other expert*' (ibid., emphasis added), in spite of the fact that the author is a respected set-theoretic topologist and his proposed refutation of the conjecture seems to deal properly with the relevant aspects of the problem. Then, the advice could be: ask him directly; if he is convinced of the correctness of his proof, try to check it yourself! Scanlon points out that the main problem in Knight's drafts is not the amount of prerequisites, but, on the contrary, the fact that 'almost everything is developed from scratch, so that one must hold the entire construction in one's mind without having the usual anchors of established theorems' (ibid.).

3.3 *The Holmes Case*

It is a long-standing open conjecture that Quine's system NF (New Foundations) of set theory is consistent relative to ZF (Zermelo–Fraenkel) set theory. Since 2010 Randall Holmes (one of the foremost scholars working on NF) has posted at his homepage (http://math.boisestate.edu/~holmes/) various versions of such relative consistency proof. He says he is fairly certain about the correctness of his proof[9]; his problem is rather to choose the proper approach (among various alternatives) to explain it in an *understandable* way. Since the author is still striving for a proper presentation of his proof, 'either waiting patiently for a final version, or reading and trying to understand a preliminary version seem to be reasonable options' (Klimpel 2015).

But the serious underlying general issue is, of course, under what conditions it can be officially claimed that a long-standing open problem in mathematics has been settled. On this point, Harvey Friedman gives the important example of what happened with Wiles's proof (1994) of Fermat's last theorem, which is, by the commonly

[9]And also that its refinement could yield the right consistency strength for NF, which is apparently much less than the strength of ZF, being that of Zermelo set theory with comprehension restricted to bounded formulas.

accepted current standard of evaluation, a conclusively proved mathematical result (Friedman 2018). First, the proof, submitted to one of the top mathematics journals, was reviewed with exceptional attention, in view of the importance of the result that was at stake: it was divided into a number of parts, read by different referees who worked separately. Secondly, a serious gap was found in the first round of review, and this is a good symptom of the care of the reviewers. Finally, the proof was fixed and published, and then it was the subject of a lot of seminars involving many specialized expert mathematicians, and during the following decades not even the slightest doubt has ever arisen.

3.4 *The Mochizuki Case*

The Mochizuki case is perhaps the most disquieting one, and certainly embarrassing. To tell the facts, we shall start from the very clear exposition by Klarreich (2018).[10] The ABC conjecture, one of the most important and difficult open problems in number theory, having a lot of consequences, says that for every $\varepsilon > 0$ there are only finitely many triples of coprime positive integers a, b, c such that $a + b = c$ and $c > d^{1+\varepsilon}$, where d denotes the product of the distinct prime factors of the product abc. In 2012, the eminent mathematician Shinichi Mochizuki in Kyoto posted online a series of long papers (a few hundred pages altogether, with a comparable amount of prerequisites) and claimed to have proved the conjecture, causing a sensation among number theorists. Many tried to understand his arguments (later also at conferences dedicated to the topic), but they were baffled by the very obscure style of the papers. The only mathematicians who vouched for the correctness of the proof were close to the author, they were unable to explain it to the others, but nobody else was able to single out a precise gap or mistake. When in 2017 a rumor (later denied) of publication (in the journal whose editor-in-chief was Mochizuki himself) circulated, many important number theorists expressed their distress: e.g., Frank Calegari in his popular blog (https://galoisrepresentations.wordpress.com) wrote that the situation looked 'a complete disaster [...]. We do now have the ridiculous situation where ABC is a theorem in Kyoto but a conjecture everywhere else', and a lot of colleagues agreed. At that point, the young Fields medalist Peter Scholze at last reported a precise point where he found a gap: the end of the proof of Corollary 3.12 in the third paper. Then the Fields medalist Shigefumi Mori in Kyoto arranged for a meeting between the opponents, which took place there in March 2018. Finally, in September of the same year, Scholze and Jakob Stix, after their long work and their visit to Mochizuki and his colleague Hoshi, posted online a report on a 'serious, unfixable gap' (at the above mentioned point) in the purported proof. The trouble is that the outcome was *a deadlock*: neither party was convinced by the arguments of the other, and the reputation of all the mathematicians involved makes the situation really

[10]The main references are given there. But see also, for a dissenting voice on the topic, Ivan Fesenko's papers: https://www.maths.nottingham.ac.uk/plp/pmzibf/nov.html.

odd. Scholze is rather skeptical and claims that Mochizuki must explain much more clearly the crucial step; Mochizuki attributes the impression of the presence of a gap to a fundamental misunderstanding of the whole work. In any case, the progress with respect to the original predicament lies in the emergence of a precise, understandable objection on a single step. Time will tell. The lesson one could perhaps learn from this strange case is twofold: first, the motivated opinion of a large majority of the experts prevails, and the burden of proof (literally, here) rests with the minority; secondly, this happens also—in fact, especially—in instances of what we have named above *untrustworthy* relations.

3.5 The Voevodsky Case

It is very interesting to read what the late Fields medalist Vladimir Voevodsky (1966–2017) had to tell about his own experience (Voevodsky 2014), not only, again, with respect to the role of computers and to the problem of checking long, difficult proofs, highlighted above, but, more deeply, as far as self-confidence—the trust a mathematician should have *in himself or herself*—or lack thereof, is concerned. First Voevodsky recalls the well-known fact that in mathematics (as anywhere) there is *an intricate web of mutual trust* on which most of original research relies, trust based, of course, on *reputation*. This has the consequence that even when doubts arise about a result obtained by an important author, often no explicit challenge is made, and yet after a few years a gap is revealed. More frequently, subtle mistakes in papers (even studied by many people in seminars, containing results used by them later) go unnoticed for years, because 'a technical argument by a trusted author, which is hard to check and looks similar to arguments known to be correct, is hardly ever checked in detail' (ibid.). Voevodsky had both experiences, they scared him, and at a point (around the end of the 1990s) they made him wonder who would ensure him that he did not make mistakes when the arguments are really hard (more than in the previous cases), require a huge effort and are very difficult to read. At that point, he put aside his usual research and 'started to think seriously about the future'. 'I didn't have the tools to explore the areas where curiosity was leading me and the areas that I considered to be of value and of interest and of beauty. So I started to look into what I could do to create such tools. And it soon became clear that the only long-term solution was somehow to make it possible for me to use computers to verify my abstract, logical, and mathematical constructions' (ibid.).

Thus, he looked for a proof assistant which could be practical to use in his own everyday mathematical life and for the specific theories in which he worked (mainly in algebraic geometry, cohomology, homotopy theory), but he did not find it among the many proof assistants then available (it was around 2000). The basic problem, he later realized, was the need for nothing less than a new system for the foundations of mathematics, since its role was not simply to be a background standard but rather a tool for reasoning in a language understandable to computers. The main problems with the usual set-theoretic foundations (say, in ZF) were the weakness of the languages of first-order theories and the inadequacy to express certain 'higher-dimensional' constructions (roughly, in the sense of higher analogues of category theory). Then, in order to overcome these obstacles, in the years 2005–2009 the system of Univalent foundations was created, with its formal deduction system CIC, parent of the system of the proof assistant Coq, widely employed nowadays (see Voevodsky 2014 and the references there).

> And I now do my mathematics with a proof assistant. I have a lot of wishes in terms of getting this proof assistant to work better, but at least I don't have to go home and worry about having made a mistake in my work. I know that if I did something, I did it, and I don't have to come back to it nor do I have to worry about my arguments being too complicated or about how to convince others that my arguments are correct. *I can just trust the computer* (ibid., emphasis added).

This is unpalatable to most mainstream mathematicians, as Voevodsky admits, but he thinks that this widespread attitude is wrong and that the original urgency he felt is still right: until proof assistants are commonly in use, mathematicians will have to face all the trouble coming from hidden mistakes and sometimes unnecessary, sometimes insufficient self-verification.

4 Some Further Philosophical Remarks

4.1 Reliability

A we have seen above, dealing with the notorious issue of the degree of reliability of the mathematical proofs that one can find in the major mathematics journals, Frans and Kosolosky (2014) consider some aspects of the actual practice in the various steps of the process of publication (peer review, revision, acceptance, critical reception by the community), showing that the naive expectation that mathematical literature is error-free or at least substantially immune from the sources of unreliability typical of other areas is a sort of wishful thinking. We have seen that they also give some suggestions to improve the practice, in terms of transparency and better credit allocation, both for authors and reviewers. But what is more interesting, from our present point of view, is again their recall of the crucial role of trust in the advancement of mathematical knowledge. They maintain (Frans–Kosolosky, 352, also citing Hardwig 1985) that trust in the testimony of others is the ultimate basis of a significant part of one's mathematical knowledge and that epistemic autonomy is an illusion: 'it would, for instance, require a lot of time and expertise for a mathematician to check the reliability of each accepted proposition in a proof' (ibid.). But what are the circumstances in which one is *justified* in relying (possibly 'blindly') on testimony, since there is no time or energy to check the work of others?

This is the object of the epistemology of testimony, but apart from the formal aspects involved (see above), it seems relevant here to recall that in fact mathematicians usually become confident that a long proof is correct only when they become familiar with the decisive new ideas that allow the proof to be carried out, and 'people are able to reconstruct the proof for themselves' (Chow 2018).[11] Thus, arguably 'the future for mathematics lies in structured and simplified proofs such as Hales's blueprint proof and in giving mainstream mathematicians the means to digest them piecemeal at their own pace and in greater numbers' (Smith 2018). The role of proof assistants (suitably streamlined and made extremely user-friendly, comparable to current web navigation interfaces, and capable to produce modular proofs) in this imagined future is, of course, crucial. This is precisely what emerges in Voevodsky's own direct testimony.

4.2 A Possible Philosophical Lesson

Let us try to collect a few remarks, although no clear, univocal philosophical lesson seems to emerge from the cases we have briefly discussed. The basic question can be formulated as follows: *what is the degree of reliability of the usual procedures which are adopted at present in order to evaluate whether very long and complex proofs of significant theorems are free from error?* A second, ensuing question is: *what are at present the possible, effective protections against the wrong acceptance by the mathematical community of incorrect proofs with the above features?*

It is not easy to give satisfactory answers, but we think there is a commonsense point on the issue which can be made (it was hinted at above). It is a fact that (as Tim Chow rightly emphasizes, see Chow 2018) in order to accept the proof of a big result mathematicians usually want to understand the reason *why* the proof works, what are the new *ideas* involved, ideas which (in all known relevant cases) were necessary to overcome the obstacles which prevented the solution, and even (as it often happened) sufficient to obtain unexpected further consequences. This process, of course, may require a rather long time.

On the other hand, this need of ideas and reasons might not belittle the role of proof assistants at all, especially if and when they are really user-friendly. If (precisely as in the Mochizuki case) a small number of mathematicians say that a certain step in a long proof, questioned by others, is trivially correct, and the step can somehow be formalized and made accessible to a suitable proof assistant (this is, on the contrary, very unclear at present in the above mentioned case, which is completely different in this respect from the Hales case), recognized as faultless by both parties, then the impasse could be broken by simply looking at the operation of the assistant (see again Chow 2018). Otherwise, as Smith (2018) complains, we would have to face situations of stalemate, in which nobody can have the last word, with a few mathematicians claiming to understand a whole huge proof (with all the connected issues) that nobody (and nothing!) else can verify.

---

[11] As Pierre Deligne once famously quipped, 'I believe in a proof if I understand it'.

5 Concluding Remarks

In our discussion of various philosophically relevant aspects of the role of trust in mathematics, in particular with respect to the notion of proof, we have taken into account some of Lakatos's reflections, the dialectics between informal and formal proofs, the recent evolution of the very concept and practice of proof, the delicate role of referees, the unavoidability of trust in any current scientific enterprise, and some attempts at formalizing the trusting relation itself. Then, we have gathered some material for case studies, on disputed (alleged) proofs, the role of computer proof assistants, the personal experience of some mathematicians relevant to our topic, and the absolute need for *ideas and reasons* in the practice of proof as a human endeavor.

Remembering Leibniz's ideal of *calculemus*, and his hope to compose in this way any dispute by assigning weights to rival theses and then calculate, we realize that certainly the role of proof assistants and automated proofs should not be overemphasized. However, if the aim is to enable people to master the general structure of a proof, then it is difficult to maintain that automated or semi-automated proofs could constitute an obstacle to the achievement of this goal. Automated proofs come under the scope of both meanings of "formal" highlighted above (Sect. 2), and may surely help to make a proof more comprehensible (Sect. 3.1) and also more reliable (Sect. 3.5). Thus, another remarkable outcome of the *formalization* process, which contributes to make the procedures of mathematical practice more and more explicit and checkable, is that a further guarantee of the soundness of the arguments involved is provided, enhancing trust.

In conclusion, although we have not obtained clear-cut answers, or advice to be given, we hope to have shown, at least, that in mathematical practice itself trust is indispensable, with all the ensuing risks, but that this fact is far from justifying pessimistic or irrationalist attitudes.

Acknowledgements We wish to thank Roberto Gronda, Paolo Mancosu and Giuseppe Primiero for helpful suggestions.

References

Alama J, Kahle R (2013) Checking proofs. In: Aberdein A, Dove IJ (eds) The argument of mathematics. Springer, Dordrecht

Bundy A et al (2005) The nature of mathematical proof. Philos Trans R Soc A 363(1835)

Chow T (2018) FOM mailing list, Jan 2018. https://cs.nyu.edu/pipermail/fom/2018-January/

Coleman E (2009) The surveyability of long proofs. Found Sci 14:27–43

Detlefsen M, Luker M (1980) The four color theorem and mathematical proof. J Philos 77:803–820

Frans J, Kosolosky L (2014) Mathematical proofs in practice: revisiting the reliability of published mathematical proofs. Theoria (Revista) 81:345–360

Friedman H (2018) FOM mailing list, Jan 2018. https://cs.nyu.edu/pipermail/fom/2018-January/

Hales T et al (2008) A special issue on formal proof. Not Am Math Soc 55(11)

Hales T et al (2017) A formal proof of the Kepler conjecture. Forum Math Π 5:1-29

Hardwig J (1985) Epistemic dependence. J Philos 82:335–349

Hardwig J (1991) The role of trust in knowledge. J Philos 88:693–708
Klarreich E (2018) Titans of mathematics clash over epic proof of ABC conjecture. Quanta Magazine, 20 Sept 2018. https://www.quantamagazine.org/titans-of-mathematics-clash-over-epic-proof-of-abc-conjecture-20180920/
Klimpel T (2015) Mathoverflow, 29 July 2015. https://mathoverflow.net/questions/132103/
Lakatos I (1976) Proofs and refutations. CUP, Cambridge
MacKenzie D (2001) Mechanizing proof. Computing, risk, and trust. The MIT Press, Cambridge, Mass
Mercier H, Sperber D (2019) The enigma of reason. A new theory of human understanding. Harvard UP, Cambridge, Mass
Primiero G, Kosolosky L (2013) The semantics of untrustworthiness. Topoi 35:253–266
Primiero G, Taddeo M (2012) A modal type theory for formalizing trusted communications. J Appl Logic 10:92–114
Ruelle D (2000) Conversations on mathematics with a visitor from outer space. In: Arnold VI et al (eds) Mathematics: frontiers and perspectives. American Mathematical Society, Providence, pp 251–260
Scanlon T (2010) Mathoverflow, 22 April 2010. https://mathoverflow.net/questions/22142/
Smith J (2018) FOM mailing list, Jan 2018. https://cs.nyu.edu/pipermail/fom/2018-January/
Voevodsky V (2014) The origins and motivations of Univalent foundations, 26 March 2014. https://www.ias.edu/ideas/2014/voevodsky-origins

Ontology, Ethics, Communication

On the Ontological Status of Trust: Robert Spaemann's Philosophy of the Person as a Promise

Walter Schweidler

Abstract Promising is not just one of many genuinely personal actions. According to Robert Spaemann's philosophy of the person, promising is the paradigmatic act in which one recognizes that one knows what it means to be a person. The ontological reason for this fundamental aspect of promise is that the being of the person is itself essentially constituted by recognition. Without recognition we would not be persons; but, at the same time, what we recognize when we act as persons is certainly a status which precedes any act of its recognition. This paradoxical constellation makes the promise a paradigmatic example for the phenomenon of fulfilled presence.

Keywords Recognition · Forgiveness · Paradox · Fulfilled presence

1 Trust Between Subjectivity and Objectivity

At first glance, trust appears as an essentially subjective phenomenon. When I tell somebody that I trust him or that I trust in God, I express an inner state—like a feeling—and a personal disposition—like an attitude. But we need only a little moment of reflection in order to notice the complex and complicated presuppositions which we make when we express our trust. As a speech act, the expression of trust resembles less an elementary illocutionary act like "I would ask you to stop smoking" or "I appoint you as commander" than a confession like "I love you" or "I believe in God". What is the point of comparison? It has to do with that very special phenomenon which we could call the experience of fulfilled presence.[1] When it is an honest act, to express my trust means to recognize what is already there when I express it. In this respect, it resembles an expression like: "I am happy". To be happy means to recognize that one is happy. There is a great variety of these kinds of phenomena. When I say "I am hungry", I recognize that my hunger is there, and this means: it has

[1] Cf. Walter Schweidler, "Vollendete Präsenz", in *La presenza, Archivio di filosofia*, ed. by Stefano Bancalari (Pisa and Rome: Fabrizio Serra editore, 2019), pp. 61–70.

W. Schweidler (✉)
Catholic University of Eichstäett-Ingolstadt, Eichstätt, Germany
e-mail: walter.schweidler@web.de

A. Fabris (ed.), *Trust*, Studies in Applied Philosophy, Epistemology and Rational Ethics 54, https://doi.org/10.1007/978-3-030-44018-3_8

been there before I recognized it. The most prominent account of this is the famous passage in Aristotle's *Metaphysics* in which he explicates the difference between *motion* and *actualization*[2]: When I am counting, I cannot say that I have already counted, and when I am learning, I cannot say that I have already learnt; but when I say that I see you in the mirror, I recognize that I have seen you already, just as when I say I "see" in a group photo in the newspaper my father, I must, in order to say this, have already discovered him. There are phenomena which are constituted by this very special temporal aspect rooted in the human condition as an *animal symbolicum*: We recognize something as existing before we recognized it but, at the same time, as something which is ontologically dependent on its recognition. If I had not recognized my father in the photo, I would not have "seen" him at all. If I had not recognized my hunger, I would not have been hungry at all; there is no unconscious hunger. And yet the act of recognition is not identical with what is recognized. When I ask the person sitting next to me on the train to watch my bag for five minutes while I go to the dining carriage to buy a cup of coffee, I must have recognized that person as trustworthy, although this "recognition" may have been a very intuitive, almost unconscious act. But whatever kind of act this may precisely have been, it was at the very least an act of acknowledgement of some objective content which is intrinsically connected with the subjective aspect of any experience of trust.

When we turn to the history of philosophy, we see that an author like Hobbes who much emphasizes the subjective aspect of confidence as a human passion nevertheless refers to its objective content as an essential condition: The joy which we call a man's "glorying" is, according to Hobbes, "if grounded upon the experience of his own former actions,…the same with *confidence*; but if grounded on the flattery of others, or only supposed by himself, for delight in the consequences of it, is called *Vainglory*".[3] Confidence, if rightly so called, requires good reasons for having it. What Hobbes indirectly confirms here is the category which in pre-modern philosophy was the almost self-evident conceptual key to the phenomenon of trust: virtue. Cicero understood self-confidence as an implication of the fundamental virtue *fortitudo*, Seneca related it to the *conscientia*, and for Aristotle trust was of course embedded in the structure of the deepest of all human relationships, i.e. friendship.[4] What is implied here is the sociocultural, institutional presupposition of any subjective experience of trust and confidence which we then find explicitly analyzed in Hegel's conception of interpersonality and in his model of "objective spirit". "Whomsoever I trust, his *certainty of himself* is for me the *certainty of myself*", so

[2]Aristotle, *Metaphysics*, in *The Complete Works of Aristotle: The Revised Oxford Translation*, ed. by Jonathan Barnes, 2 vols (Princeton, NJ: Princeton University Press, 1995), II, bk 10, Chap. 3, 1047 ff.

[3]Thomas Hobbes, *Leviathan: with selected variants from the Latin edition of 1668*, ed. by Edwin Curley (Indianapolis, IN and Cambridge: Hackett, 1994), Part 1, Chap. 6, § 39.

[4]Cf. Tanja Gloyna, "Vertrauen", in *Historisches Wörterbuch der Philosophie*, ed. by J. Ritter, K. Gründer and G. Gabriel, 13 vols (Darmstadt: Wissenschaftliche Buchgesellschaft, 2001), XI, pp. 986–990.

the *Phenomenology of Spirit*[5]; and "objective spirit" is Hegel's designation for the sociopolitical structure which ontologically, not in any purely ethical sense, depends on trust. A state exists only when its laws are obeyed, but the obedience to the laws presupposes that its citizens trust in the order and thereby in the existence of the state. Let me quote the most relevant passage of the *Rechtsphilosophie*: "The political disposition, i.e. patriotism in general, is certainty based on truth (whereas merely subjective certainty does not originate in truth, but is only opinion) and a volition which has become habitual. As such, it is merely a consequence of the institutions within the state, a consequence in which rationality is actually present, just as rationality receives its practical application through action in conformity with the state's institutions. —This disposition is in general one of trust (which may pass over into more or less educated insight), or the consciousness that my substantial and particular interest is preserved and contained in the interest and end of an other (in this case, the state), and in the latter's relation to me as an individual. As a result, this other immediately ceases to be an other for me, and in my consciousness of this, I am free".[6] Structures of this kind, although depending on subjective consent, are real in a higher sense than any empirical facticity, exactly because they rest on an interpersonal relation; a community into which the individual is already integrated when it comes to his or her consciousness that this is so. They are, we could say, the interpersonal, sociocultural variation of the phenomenon of fulfilled presence.

It is this essentially temporally constituted structure which marks the starting point for Robert Spaemann's philosophy of the human person as a promise.[7] His general idea is to anchor the interpersonal structure of mutual recognition in the self-relation of the individual person. The connection between these two relational realities is for Spaemann the key to the understanding of human dignity. Human dignity is not something that could be first discovered and then respected; rather, both the claim to and the respect of human dignity are, to use a Heideggerian term, equiprimordial. The question in what another person's dignity consists cannot be answered by description but only by action. But action presupposes what it at the same time constitutes: recognition. For the agent who has to respect all other human beings as persons, there is no remainder of receptivity which would precede the cultural form of life that constitutes the unity of humanity. "This, of course, is paradoxical. Respect, recognition, and so on are species of activity, which would seem to presuppose a moment of receptivity in which persons were identified as persons. Apparently, and not least because it is a question of perceiving another center of being, the perceiver needs to be in a completely receptive posture. But that is not the case, and for a very clear reason: A center of being is, by definition, not something available to knowledge

[5]G. W. F. Hegel, *The Phenomenology of Spirit*, trans. by A. V. Miller (Oxford: Clarendon Press, 1977), § 549.

[6]G. W. F. Hegel, *Elements of the Philosophy of Right*, ed. by Allen W. Wood, trans. by H. B. Nisbet (Cambridge: Cambridge University Press, 1991), § 268.

[7]See Robert Spaemann, *Persons: The Difference Between "Someone" and "Something"*, trans. by Oliver O'Donovan (Oxford: Oxford University Press, 2006).

as a phenomenon".[8] In this constellation we face the deepest and genuinely temporal core of the paradox of our human existence: that in some way our existence is a response that precedes the question to which it is the answer. And the genuinely challenging aspect of this paradoxical constellation consists in the fact that we as human beings cannot but rely on the understanding of this paradox and the obedience to it. To "rely" here means of course: to act in a way which implies and demands of others to confirm and constitute the reason which we must already rely on when we act. And in this sense, the claim to recognize human dignity is "agent-relative"[9]: The universal relation that connects all of us as members of the human family intrinsically presupposes an acting subject who understands that in the act by which he or she recognizes it nothing less than that universal relation itself is at stake. If we want to give a philosophical account of this, we cannot, according to Spaemann, refer to any normative principles, but rather we have to turn to fundamental phenomena which incorporate the solution to our problem. Promise is one of them.

2 The Person as Promise

The starting point is to direct one's attention to the fact that only persons can promise. This is much more than, for example, to remind oneself that only persons can play chess or make chairs. The promise is a sociocultural institution which represents in a metonymical way the whole human existence as an *animal symbolicum*, i.e. it is only one among all the other typically human forms of social communication but at the same time it represents something that connects them all. The barkeeper mixes a good drink and serves it because he relies on the guest's understanding of and will to respect the obligation to pay. He does not ask where this disposition comes from; he simply trusts the other person. And the same with the guest, when he pays: He fulfills an expectation for the reasons of which he does not ask. Actually, the whole practice would be sensitively disturbed if he, when ordering the drink, should say: "You can count on me to pay..." When we make a promise it is different. Of course there are many social and cultural conditions of performing this action which we do not bring to our mind at the moment of promising; the origin of the whole practice is not and could never be completely transparent for us. But there is one thing which is typical for what I called the metonymic aspect of the action named promise: When I make a promise I put myself (and the one to whom I make the promise) in this origin. I quasi attend here and now the event that takes place whenever and wherever a promise is made. This metonymical aspect is decisive for the genuine importance of the institution called promise, but also for the concrete practical consequences it has for anybody who participates in it. When we promise something, we know "that what we base our expectation on is precisely the promise. Promises create and justify

[8] Ibid., p. 182.

[9] Cf. Thomas Nagel, *The View from Nowhere* (Oxford: Oxford University Press, 1986), pp. 164 ff.

expectations prima facie, because they give rise to claims".[10] He who reminds me of my obligation to fulfill what I promised does not need to repeat the reasons why I gave it to him, he just needs to remind me of the promise itself. What does this show us? It shows that the real core, the ontological substance of the sociocultural practice called promise is to be found in the *person* who promises. "A promise licenses a claim that is peculiarly unconditional; but if our expectation of its fulfilment is not quite so unconditional, that is because we do not know how well any given individual will measure up to its unconditionality. Moral necessity is not physical necessity. Whether it is compelling as physical necessity depends on who it is that undertakes the obligation".[11] So, coming back to our remark that, when we make a promise, we somewhat attend the origin of the institution itself and the condition of its possibility, we can now even say that in the event of promising we touch the origin of our own and, given the metonymical perspective mentioned above, even that of everyone elses' personality itself.

Here we have reached the point in which Spaemann confronts us with the thesis that the person in itself is a promise. What is the philosophical *pointe* of this thesis; the general idea which makes it more than a rhetorical summarization of what we said up to now? The decisive point is that this thesis is a statement of the sort to which Wittgenstein referred when he ascribed fundamental philosophical relevance to Goethe's term *Urphänomen*[12]: The spade bends back, the fact brings our reflection to an end, the description releases us from any further search for explanation, we are as certain of our insight as we are of our sight and of the meaning of the words we use to express it. I shall quote a longer passage from Spaemann because it is fundamental for the methodological core of his whole conception of personality. In this passage, he responds to the question why we cannot strengthen or support a given promise by promising to keep it: The fact of the promise itself precedes any possible normative confirmation which we could try to find for it. When Spaemann uses here the expression "seems to", we must think of the decisive Husserlian-Heideggerian device: "So much semblance, so much being".[13] "The distinctive feature of moral obligation seems to be that it rules out a reflective train of thought that is in itself perfectly possible, by which we may wriggle out of any obligation at all. To renounce this train of thought seems to be the very essence of the moral act. In renouncing it, the human subject realizes him- or herself as a person, and that is the irreducible presupposition of the train of thought itself. One takes upon oneself the promise which, as a person, one already *is*. By speaking and demanding to be understood, one has engaged in the same personal relation that is presumed in each separate act of promising. The question of how to secure the promise no longer arises. The final

[10]Spaemann, *Persons*, p. 221.

[11]Ibid., p. 222.

[12]Cf. Ludwig Wittgenstein, *Philosophische Untersuchungen/Philosophical Investigations*, trans. by G. E. M. Anscombe, P. M. S. Hacker and Joachim Schulte, rev. 4th ed. (Oxford: Blackwell, 2009), § 654. One can also think of Jean-Luc Marion's term *phénomène saturé*.

[13]Martin Heidegger, *Being and Time*, trans. by John Macquarrie and Edward Robinson (Oxford: Basil Blackwell, 1985), Rzn. 36.

security is the refusal to pose that question, a refusal already made when human beings recognize each other, as persons. The person is a promise. To be a person means to assume a place in the community of all persons. Intelligent creatures have no choice whether or not to assume this place together with its associated obligations and claims. They can neither refuse the claims nor renounce the obligations. The person is autonomous; but there is no autonomous act giving rise to personal autonomy, no autonomous discretion to which that act is subject".[14]

One takes upon oneself what one already is but what, at the same time, one essentially becomes by taking it upon oneself: Here we face again the genuinely temporal structure of fulfilled presence. My being as a person depends on a decision the content of which is prior to itself and yet, at the same time, is taking place when I make a promise. A person's being contains in itself a structure of the repetition of an origin which is constituted by that repetition.[15] It may sound like a sort of psychoanalysis, but, if this is so, it only shows us that psychoanalysis is itself rooted in that ontological structure which, according to Spaemann, has its real foundation in the relation between the being of the person and its essential substance, i.e. *human nature*. To understand this, we must of course understand the term "nature" in the sense in which it was conceived by the metaphysical tradition, namely as a concept of distinction, not reduction. Its nature is what distinguishes a living being from beings of another species, and in this sense of course there is a human nature which shows itself in the forms of life that characterise the human in difference to all other natural beings. Everyone knows the features which the metaphysical tradition ascribed to that nature and the forms of life in which it shows it comes into appearance: *zoon politikon*, *zoon logon echon*, *animal rationale*, *imago Dei* etc. What we, according to Spaemann, learn furthermore about ourselves when we turn our philosophical attention to the phenomenon of promise, is that there is not only that natural difference to all other natural beings; it is something additional. What distinguishes us from all other natural beings more than our nature as such is the relation in which we stand toward this nature, and this is the relation of freedom. When we "follow" our nature, we do it by free decision, and we can always refuse to do it. Therefore, human beings face the choice between human and therefore humane and inhuman and therefore inhumane action. This is the reason why our respect for the dignity of other human beings cannot be explained as a form of "speciesism". A human being can treat his own kind in ways which are incomparable to any other species: torture, concentration camps, mass destruction etc. If we forbid ourselves from acting in such an inhumane way, we follow our nature, but we do this, like no other natural being, on the basis of free decision. Freedom, however, must be understood essentially from the standpoint of the paradox which we faced in the aspect of the self-repeating origin of our personality: The freedom that constitutes our dignity as persons is,

[14] Spaemann, *Persons*, pp. 222 f.

[15] Cf. Walter Schweidler, "The Self-Repeating Origin: Ontological Aspects of Ricœur's Concept of Hermeneutics", in *Hermeneutics and the Philosophy of Religion: The Legacy of Paul Ricœur*, Claremont Studies in the Philosophy of Religion, Conference 2013, ed. by Ingolf U. Dalferth and Marlene A. Block (Tübingen: Mohr-Siebechk, 2015), pp. 81–96.

according to Spaemann, rooted in the fundamental circumstance that, different from any other natural beings, we are not simply what we are by nature rather we *have* our nature. "Freedom", so he writes, "is first and foremost freedom *from* something; but what is the *person* free from? Only from his or her own nature. A person 'has' a nature, but the nature is not what the person *is*, because the person has the power to relate freely to it. But this power is not innate; it comes through encounter with other persons. Only the affirmation of other centers of being, through recognition, justice, and love, allows us the distance on ourselves and the appropriation of ourselves that is constitutive for persons—in sum, 'freedom from self'. This we experience as a gift".[16] When we finally try to explicate what this means: that we "have" our nature, the special importance of trust for our human condition can be pointed out.

3 To Thine Own Self be True

Trust simply is the subjective side of the objective content and substance of the promise which the human person *is*. With this remark we come back to what we mentioned regarding Hobbes and the pre-modern philosophers. The relation that constitutes the human person is essentially a relation *between* persons. The concept of person, according to Spaemann, necessarily presupposes a plurality of persons.[17] In fact, it necessarily implies the totality of persons, i.e. we devaluate and destroy the personal relation whenever we exclude any being who has a human nature from the community of persons.[18] "There is no act of co-optation, but everyone takes a place in this community by birth-right…The original promise undergirding all subsequent promises stands beyond the reach of question or challenge, simply because it was never a free action performed in time. It is an intelligible postulate, the meaning of which is nothing more or less than our actual situation in the community of persons".[19] Insofar and in this respect Spaemann obviously opts for the Kantian position, and he emphasizes Kant's parallelization between the normative constitution of the community of persons and the natural laws: "Act as if the maxim of your action were to become by your will a universal law of nature".[20] The decisive point for Spaemann here is the "as if" which shows us: "We cannot will this universal law to be a new law of nature in fact, because we know that is impossible…The maxim is not to become a law of nature, but to be as much like one as possible *in respect of its regularity*.

[16]Spaemann, *Persons*, p. 216.

[17]See ibid., p. 27; "Persons exist only in the plural" (ibid., p. 232).

[18]The "relational sphere of personal interaction is universal, from which it follows that the exclusion of even one person from the scheme of recognition brings the personal character of the whole system tumbling down" (ibid., p. 196).

[19]Ibid., p. 223.

[20]Immanuel Kant, *Groundwork of the Metaphysics of Morals*, ed. and trans. by Mary Gregor (Cambridge: Cambridge University Press, 2006), p. 31.

The one serves as a model for the other".[21] I will come back to the very last point of this quotation in the end; but for now, the decisive aspect of the "as if" is the hidden temporal condition of the concretization of that "intelligible postulate" in our real world. This is exactly the point where for Spaemann trust enters the game.

Regularity, according to Spaemann, "has two roots. Its specific and peculiar root is personal trust, the faith reposed in the word and free commitment of the promise. The object of this trust is the other's freedom, and trust is clearer and stronger when backed by the conviction that the other was really free from impulsive inclinations, independent of natural determinants, 'master of himself'. The more the will is independent of nature, the more it corresponds to the model of a law of nature".[22] This paradox is the key to the other root which concerns our very particular, our individual subjectivity and from which we come to know what is the practical consequence of our insight into the promise we *are*. "In making a promise one must…mistrust oneself. In making a promise one must be willing to cultivate in oneself the right inclinations to support promise-keeping. One must assent, furthermore, to the legal sanctions that make breach of promise disadvantageous. They will not disadvantage oneself unless one intends to break one's promise. But the knowledge that they *could* affect oneself helps one what one anyway intends to do simply for the promise's sake".[23] So, the fundamental practical conclusion from the transcendental insight into man's existence as being a promise is *the work on one's own trustworthiness*. In fact, this can be seen as the ethical principle in which Spaemann has found a kind of reconciliation of metaphysics and phenomenology. On the one side, the person, as a center of being, is not itself a phenomenon.[24] But on the other side, the relation which allows one to direct one's life according to that transphenomenal core of one's own being is not primarily the relation to an extratemporal, ideal source of normativity but it is a structure which is present in our ordinary life and especially in the everyday experience of personal trust, metonymically incorporated in the practice of making and obeying one's promises. There is a doubling of oneself which is not primarily grounded in the relation which defined the Kantian concept of person: the relation between the animalic natural individuality and the superindividual transcendental self; but it is grounded in the paradox of finding oneself by means of getting into distance to oneself. One point of reference for Spaemann has often been Harry G. Frankfurt's concept of "second order volitions".[25] But we can trace back his sources to Aristotle, who in his *Politics* made the distinction between the "tyrannical" rule of the soul over the body and its presupposition, i.e. the not tyrannical but "royal" or "political" rule of reason over the desires.[26] To become the master of oneself not, like

---

[21] Spaemann, *Persons*, p. 230. The translation "regularity" is not completely free from doubt; the German original is "*Verläßlichkeit*" (reliability).

[22] Ibid., p. 230.

[23] Ibid., p. 231.

[24] Cf. the quotation above, fn. 8.

[25] See Harry G. Frankfurt, "Freedom of the Will and the Concept of Person", *Journal of Philosophy*, 68. 1 (1971), 5–20.

[26] Cf. Thomas Buchheim: "Wie die Vernunft uns handeln macht", in *Die Normativität des Wirklichen: Über die Grenze zwischen Sein und Sollen*, ed. by Th. Buchheim, R. Schönberger and

for Kant, by listening to some transcendental master's voice, but by developing the confidence to attend and accept the fulfillment of the promise which one's life means for itself: This is the personal solution to the problem of the paradox of human freedom. This is the point where for Spaemann the metaphysical tradition has supplied us with the last word that can be said about the mastery over oneself: conscience. "To make conscience a reliable factor—and in cases of conflict, a conclusive factor—in one's psychological make-up, is a task only beings with 'secondary volitions' can confront, those who can take a view of their impulses as a whole, i.e. persons. The purpose of this effort is what we call 'virtue': conditioning one's nature to reliable self-determination, integrating the various impulses with the goal of actually achieving what one wishes to achieve. What is at stake is the ability to rely on oneself".[27] There can be no philosophical understanding of such a device without going into and analyzing the deeply temporal presuppositions which it has. The work on one's life is more than and something categorically different from drawing the balance of one's achievements relative to the expectations one has once had. It is a work on the mutual illumination of expectation and fulfillment; it requires an—maybe unphilosophical or at least pre-philosophical—understanding of the phenomenon of fulfilled presence.[28]

In the phenomenological view, there cannot be a principle of deduction for the validity of this analysis of promise and trust, but only a kind of confirmation by the demonstration of another phenomenon which is essentially connected with promise and decisive for its importance for the human being. That "co-phenomenon" of the promise is for Spaemann *forgiveness*. It is the ethical institution which makes possible to make even the broken promise a sign of humanity. "One who forgives abandons the right to take the offender as he found him, and gives the offender the opportunity to takes himself differently, too."[29] The paradox is here that someone can be more true to another's self than to his own. The wrong action cannot be undone, but it can be taken as the chance to confirm the truth. "Promise underwrites the independence of personal identity from absorption in actuality. Forgiveness evokes this independence in the teeth of actuality. It is, in an eminent degree, an act of creation."[30] These are Spaemann's last words in his book on *Persons*.[31] In his book *Happiness and Benevolence* from 1996 he has gone a little step further. There he

W. Schweidler (Stuttgart: Klett-Cotta, 2002), pp. 281–413 and Walter Schweidler: "The Slave in Ourselves. On the Importance of the Aristotelian Concept of Slavery for Political Legitimation", *Working Papers in Philosophy*, (2013/2014), 1–17. https://fi.btk.mta.hu/images/Események/2013/working_papers/2013_04_walter_schweidler_the_slave_in_ourselves_on_the_importance_of_the_aristotelian_concept_of_slavery_for_political_legitimation.pdf.

27 Spaemann, *Persons*, p. 225.

28 Cf. Walter Schweidler, *Geistesmacht und Menschenrecht: Der Universalanspruch der Menschenrechte und das Problem der Ersten Philosophie* (Freiburg and Munich: Alber, 1994).

29 Spaemann: *Persons*, p. 234.

30 Ibid. p. 235.

31 The final chapter. "Are All Human Beings Persons?" was added to the text; it comprises Spaemann's lecture given on the occasion of the bestowal upon him of an honorary doctoral degree at the Catholic University of Washington in 1995.

raises the question if there is something like forgiveness in relation to oneself. It is, according to Spaemann here, a fact that in certain life situations we have the feeling of a responsibility to ourselves and for the life we have to live. The philosophical account of this feeling cannot be based on the Lockean idea of self-possession. If I were the possessor of myself, I would have to presuppose an act of appropriation; nobody possesses anything without, at least impliedly, seizing it. But who would be the subject of such an act of appropriation? If we answer this question with "myself", we tumble into a *regressus ad infinitum*. So, the only alternative for Spaemann is "that we think of the human as an image and a representative of an unconditioned, that is, if we think of the human as someone who does not belong to him or herself." And thereby we come to think of ourselves to be responsible for ourselves, which is paradigmatically a religious thought. "The thought of responsibility for oneself is a religious thought; it is even a constitutive thought for religion."[32] And as such a thought, it requires the assumption of a person prior to anyone who may ask for forgiveness. "The thought of responsibility to oneself can only be consistently held when forgiveness is also a possibility, and this possibly includes a subject of forgiveness."[33] Spaemann renounced what he originally wanted to finally add, i.e. the sentence: *Et hoc est quod omnes dicunt Deum*. But this sentence clearly formulates a presupposition which we will have to make and to discuss when we speak about his idea of the human person as a promise. This discussion could bring us to a very special point, i.e. the reactualization of a seemingly outdated belief: the genuinely *cultic* self-understanding of the human being as being responsible for the continuation of the life of nature.[34]

[32]Robert Spaemann: *Happiness and Benevolence*, trans. by Jeremiah Alberg (South Bend, in: Notre Dame Press, 2000), p. 198.

[33]Ibid., p. 199.

[34]Cf. Walter Schweidler: "On the Social Origin of Time in Language", in *Origins and Futures: Time Inflected and Reflected*, ed. by Raji Steineck and Claudia Clausius (Leiden and Boston, MA: Brill, 2013), pp. 37–48.

Can We Trust Machines? The Role of Trust in Technological Environments

Adriano Fabris

Abstract In this paper, I will deal with trust in technological devices. Nowadays, we are led to trust AI machines, extending the notion of "trust" to our relationship with such devices. After reviewing different forms of trust and showing that trust is inextricably linked with a sort of fear, this paper will try to relate all this to the ethical judgement we can form about what machines "do" as well as about the machines themselves. Prompted by this, the paper will discuss the case of a recent attempt made by the EU at working out ethical guidelines for a trustworthy AI. To conclude, it will try to answer the above question—can we trust machines?–and accurately define on what terms they can be trusted.

Keywords Trust · Machine · AI · Technology · Ethics

1 The Word "Trust"

Can we really trust machines? Can we trust something artificial? Why can we rely on what has been built by us but that, in many cases, seems to be able to "act" independently of us, since it has a sort of "intelligence", Artificial Intelligence (AI)? How is the notion of "trust" affected by these new relations? These are the questions I want to try to answer in this paper.

To do this, the terms of the question must first be explained: literally. We must actually perfectly understand the meaning of the words we use. I am speaking of the terms that, in European languages, are associated with the semantic dimension of "trust". I mean the English "trust" and the German "*Vertrauen*". I also mean those words that in the Romance languages come from the Latin "*fides*" and its etymology: such as the Italian "*fiducia*", the French "*confiance*", the Spanish "*confianza*", among others, as well as the English "*confidence*".

Let's start with Latin. The words "*fiducia*", "*confiance*", "*confianza*", "*confidence*" as well as their verbs are all derived from the Latin "*fides*" (De Vaan 2008, 218–219). Then, etymologically, the word "*fides*" is related to the Indo–European root

A. Fabris (✉)
University of Pisa, Pisa, Italy
e-mail: adriano.fabris@unipi.it

A. Fabris (ed.), *Trust*, Studies in Applied Philosophy, Epistemology and Rational Ethics 54, https://doi.org/10.1007/978-3-030-44018-3_9

"**bheidh/bhidh*", which can be found in the Greek verb πείθω ("I persuade", middle πείθομαι "I am persuaded" and therefore "I obey") and in the noun πίστις, "trust", as well as "faith". Here, the connection with others words, which are used in a strictly religious context, is clear. I am speaking of the words "*fede*" (Italian), the Spanish "*fé*", the French "*foi*" and, through French, the English "*faith*".

What does this far–ranging conceptual constellation, which recalls both the daily dimension of belief and the religious one of faith, mean? It generally means a gesture of openness to others. Whether it is trust in other human beings or faith in a personal God, in either case all such words express an availability to embrace what the other can give us. In trusting others, we are willing not to consider ourselves as the only guarantors of the legitimacy of some notion or the effectiveness of some action. This happens because we admit—more or less consciously—that we have limits, that we do not know everything, that we do not control everything. My turning to someone else means that I cannot get something without the other's help. So, I must trust others, I must rely on them, I must ask for their support. But first and foremost I must confide in the fact that this support will come and will actually take place in the way I expect it to: even if that's not a foregone conclusion.

The reference to the etymology of "trust" and "*Vertrauen*" shows us instead other dimensions of such reliance. Such words have the same root as "true". "True" should however be meant as "honest", "steady", not in the sense of "corresponding to facts" (as in the definition of "truth" that since Aristotle onwards has been recurring all through the tradition of Western philosophy). The common root of these words—"trust" and "*Vertrauen*"—is actually the Proto–Germanic word "**trewwu*", meaning "loyal, trustworthy", which goes back to "**trewa*": tree (Kroonen 2013, 522–23). The image suggested by such words is an image of steadfastness, reliability, firmness, which are typical of a tree trunk.

In this case, then, it is not so much our need to open up to others, to rely on what does not depend on us because of our limitations, but it is the reason for, the cause of our reliance that is highlighted here. Such reliance is owed to a quality that is specific to the one we rely on: the fact that this person or this thing deserves to be trusted by us, as they turn out to be a firm point of reference for our actions. Only because of this can the person or thing we place our trust in be considered worthy of it.

In a nutshell, the etymology of the words that are derived from Latin and those that are derived from the Proto–Germanic languages sheds light on two issues that are interconnected or actually complementary, in defining a confident attitude. The etymology that refers to the Latin "*fides*" tells us what happens, when, as we become aware of our limits, we confide in a dimension that is other than us. On the contrary, the etymology that recalls firmness, steadfastness tells us why we resort to someone or something else and trust them. We trust someone or something that provides a firm point of reference: a point of reference that we acknowledge. In this case, we do not have to sacrifice anything. Quite the opposite, what we find here is the benefit—benefiting from a firm point of reference—that we get by confiding in something else and that this confidence inspires. In other words: while the etymology of the Latin and Romance words shows us *how* trust arises, the etymology of the English and German words suggests *why* it can arise.

2 Trusting Robots

The problem we have to address now is whether such bond of trust in the two forms I mentioned before may also be formed with a machine. It looks very much like it. It is typical of our age. We trust software or applications—for instance Google Maps—sometimes even against our personal experience. We trust technological devices—such as a computer, our smartphone, and soon we will be ready to trust a self-driving car—without much thought (Botsman 2018).

But once again we have to understand how it happens, how we get to trust a machine or a programme, and even more importantly we have to understand why we do that. It is an extension of our tendency to trust. What is actually at stake here is an amplification of the terms that we can put into a bond of trust: this no longer involves human beings, some kind of animals or may be a personal God alone, now artificial devices are included too.

To understand the reasons why we extend our trust, first we need to be familiar with the artificial devices that can be the recipients of our trust today. Here I am speaking in particular of some special machines: robots. Robots are regarded both as machines that can be used to increase our power and the scope of our actions, and as systems with some degree of autonomy and, as such, capable of "acting" even without direct human input. They may even "act" more powerfully and effectively than human beings can. This is the decisive argument in the matter we are discussing here. Robots are not mere *technical* tools, which can of course increase the possibilities of human action and which cannot however work without such action, without more or less immediate human inputs. They are instead *technological* equipment, which means that, once built and programmed, they can work in their environment with some degree of autonomy. While a technical tool should always be viewed as a means to a purpose, and the way such purpose is achieved is established by the user of the tool, technological equipment can instead pursue its purposes—the ones it has been programmed for—by acting in a way that makes it effectively achieve it, without having to be constantly controlled by the user (Fabris 2018, Chap. 1). That's why the ways it can follow are many and unpredictable, depending on its interactions with the environment and the ensuing adjustment of its procedures.

Then, for such results to be achieved, many requirements need to be fulfilled: both technical and technological. A technical tool can be something isolated or something that is synergistically connected with other tools. In either case, however, it cannot work without an external action or an action that is outside of its own action. An example of the former is when you use a stick to shake fruit from a branch that is too high to be reached by a human being. An example of the latter is a mechanical clock, built with different pieces linked together (levers, springs, cogged wheels, etc.) that move a mechanism that cannot work if it is not wound by someone.

Technological equipment, instead, not only joins all such tools into a system, but, as I said, it can also "act" autonomously: at least in a specific sense of the word "autonomy" and in varying degrees. In other words, it is designed to adjust its actions by responding to feedback and correcting its actions accordingly, since, as I said, it interacts with its environment and reflects its changes, more or less accurately. Let's think of the little robotic lawn mower, which changes direction when it finds an obstacle. An example is the more complex system of a self–driving car, which adjusts itself to the movement of the other cars and beings it can meet on its way.

What's new about technological equipment is the fact that, because it can respond to environmental inputs and adjust its outputs accordingly, it can be considered, to a certain extent, an "intelligent" agent. It is in this sense that "Artificial Intelligence" (AI) has been so widely spoken about for some time now. Maybe that is the feature that can make us understand why we can have a bond of trust with technological systems and specifically with robots. What I mean is precisely the fact that robots too can have a form of "intelligence", and such "intelligence", with its distinctive features, is something human intelligence can relate to.

But what does"intelligence" mean when we have to do with robots? A recent document issued by the European Union helps us define this word. It's worth carefully reading what it says about AI: "Artificial Intelligence (AI) refers to systems that display intelligent behaviour by analysing their environment and taking actions—with some degree of autonomy—to achieve specific goals. AI–based systems can be purely software–based, acting in the virtual world (e.g. voice assistants, image analysis software, search engines, speech and face recognition systems) or AI can be embedded in hardware devices (e.g. advanced robots, autonomous cars, drones or Internet of Things applications)". Such systems display their intelligence—actually, their kind of "rationality"—"by: perceiving the environment in which the system is immersed through some sensors, thus collecting and interpreting data, reasoning on what is perceived or processing the information derived from this data, deciding what the best action is, and then acting accordingly, through some actuators, thus possibly modifying the environment" (High-Level Expert Group on Artificial Intelligence 2019a).

3 The Missing Quotation Marks and the Implication of Trust and Fear

As we can see, it is a quite thorough definition. It clearly describes the circular relationship that exists between a given kind of programmes and the environment with which they can interact. To do this, such programmes need to be embedded in a physical structure. An Artificial Intelligence that is embedded in a machine and that, through it, can work in its environment is precisely that of a robot.

At the same time, though, the definition of "Artificial Intelligence" that I quoted is also a clear sign of a linguistic conundrum or at least an improper use of the language. Such definition uses words that are normally attributed to human action and to forms of intelligence that are specific to humans. As we saw, as to systems with an "Artificial Intelligence", the definition mentions their ability to "perceive", "interpret", "reason", "decide", "learn", "analyse" their environment and "act accordingly".

These verbs seem to have been used literally, because the quotation marks that should be used to emphasise the difference between the way humans are intelligent and act, and the way artificial systems are "intelligent" and "act", are missing there. Actually, it is not the same thing. Machines "act" and "behave" according to the procedures they have been programmed for. Their actions consist in executing a sequence of predefined operations, which can still involve some implementation or adjustment, based on an interaction with the environment. Conversely, human action is not something merely procedural. It stands out because it enacts multiple forms of rationality—strategic, instrumental, emotional, etc.—that enable it to handle the contingency of the events, not only by trying to foresee and control it according to specific procedures, but also by adapting and responding to it creatively. By this, I mean that human beings can instantly change perspective when new general conditions arise and can make those leaps in their logical and decision–making processes based on the unpredictability of the events that are not covered by any procedure or any method.

In other words, while Artificial Intelligence is practiced through predictions and quick calculations, and needs to consider all reality as something completely predictable and calculable, that is, something that can be artificially reconstructed, human intelligence sees the world not just like that but above all as an occasionally unpredictable response to events that can happen. In a nutshell, human actions can of course be viewed in procedural terms as well, but that is just a view: a simplistic view of the multiple forms and different processes that it embodies.

If quotation marks disappear, that is, if no difference is highlighted between the same words used in the case of the human being and in the case of the machine, the consequences can be really remarkable, both anthropologically (about the way humans perceive themselves) and ethically (the way humans perceive their relations with other agents, including artificial ones). Such consequences are remarkable mainly for the issue I am dealing with: whether to trust an artificial device. If we actually think that robots do the same things we humans do, and in the same way as we do them, or actually better than we do them; if we think that they are somehow like us, or actually better than we are, then of course we can trust such devices. On the one hand, AI robots look familiar to us, because they "behave" like us, and, on the other hand, because in some respects they can "act" better than us, they provide us with a safe point of reference for our actions. They are like us, because they can interact with their environment. They can be held to be better than us, since their predictive abilities are wider and faster and therefore they perform better.

So, AI robots meet both requirements to be trusted, the requirements found by the previous etymological analysis. The consequence is that, because of that specific reliability and safety that we can attribute to such devices, we can even create a bond of trust with them. Yet, if you look into it, it is not just a feeling of confidence that is aroused in our relationship with such systems. Another attitude is very often aroused, which is linked with trust: an attitude of mistrust, of fear.

Robots—regardless of the etymology of their name, which as we know comes from the Czech "robota" ("servant", "slave")—cannot be regarded as mere systems at the service of human beings. Indeed: in many cases they have replaced humans now. They can perform some operations faster and more accurately than humans, and so they can replace them.

All this is very anxiety–inducing. We are concerned that robots may supplant us in the activities we do. We fear that, as a consequence of their increasing ability to interact with the environment in fairly unpredictable ways, they can get out of hands. We are upset at the fact they are very often built with human features. Actually, if robots are made in our likeness, if they look like our "lookalikes", all the more reason for seeing them as our competitors, definitely not as our partners.

Multiple reasons are interwoven here that can arouse a feeling of mistrust and a fear for machines. I'll say it again, it is a fear of replacement: that is, humans being replaced by artificial systems. It is a fear aroused by our loss of control: since the robots' increasing efficiency could not only emphasise our limits, it could even humiliate us. Then, there is a more psychological fear, a primordial fear, the one we feel every time we meet our "lookalike", a *Doppelgänger*, as the Germans call it: someone who looks too much like us. This is what has been reported, precisely in the case of our relationship with robots, by Mori Masahiro in a well–known article, explaining that the more an artificial system looks like a human being, the more "uncanny" it is felt to be by a human being (Mori 1970, 33–35).

4 Another Dimension in the Bond of Trust Between Humans and Machines

There is another side of the coin, in the two interconnected dimensions of trust that I have pointed out so far—trust that comes from my opening up to something else because I am aware of my limits, and trust that finds a sound, firm point of reference in something else that deserves to be trusted. It is the connection between trust and fear. Such connection is quite obvious even in our relationship with machines. Even if we trust a device, we cannot help thinking that, after all, it could betray us, that it will behave unpredictably, that it can break down. This happens in human relationships as well. As a matter of fact, trust and mistrust are always interwoven.

For instance, when I recognise that I am more limited than something else and confidently open up to it, I always fear this something else can take advantage of me and take the upper hand. When I realise that some devices, some phenomena or some events, even if produced by humans, can get out of human control because they have a certain degree of autonomy, then I cannot but be afraid of their unpredictability. When I see someone or something else that looks too much like me, then I can no longer identify with this someone or something else, I can't see myself in this someone or something else: just because there isn't someone or something else, there is just a "me" who isn't me. In other words, too much unfamiliarity and too much similarity can both be "uncanny", because they disrupt the chance of establishing a proper relationship.

So, trust and fear, even in relationships between humans and machines, cannot be separated from each other. We have to keep this in mind: because there is always an undercurrent of anxiety in trusting a machine, just like in trusting a human, there are always setbacks and afterthoughts. That's why technological Luddism and successful integration, apocalyptic scenarios and heavenly prospects, for instance those inspired by the post–human, live together in our age. An example is the schizophrenic attitude—sometimes categorical condemnation, sometimes compulsive use—that most people have towards their smartphones.

However, there is still one more dimension in our trust of machines that, for the sake of completeness, we must analyse. It is not only trust, as we have seen so far, or confidence, often going hand in hand with fear and caution. It is a different experience that stems from a feature of the machine and that, once recognised, can arouse trust and confidence in us. I am speaking of the meaning of the word "reliability": namely, something that is not specifically trustworthy, i.e. worth of trust, but that can provide or ensure reliance.

The word "reliance" suggests a dependency, a connection, something that fits a broader relational context (Hoad 2003). "Reliable" comes from the verb "to rely", from the Latin "*religare*" (to tie, to bind). It is the same verb from which Lactantius, in his *Divinae Institutiones*, derives the word "*religio*", religion (Lactantius 2004).

In this case, the reason for the trust and trustworthiness lies in the fact that the element at issue can be considered safe, as such: both because it is bound to other elements, and it takes its place and firmness from such bond, and because such element is structured in a steady, well defined way itself. Here, then, the reliability is due to something that fits a given system or a given procedure. And there's more. It is due to the fact that this something looks like a well–organised system itself or like a procedure that, if properly performed, will spring no surprises on us. That firmness, that steadiness that can inspire confidence depend therefore on the predictability of some procedure. That's why something can be considered to be "reliable" and one can fall back on to it.

Of course, this is a specific way in which machines and their programmes inspire confidence. Technological devices are reliable because they cause no surprises, at least as long as they work. And they cause no surprises because, when they do work, they follow specific procedures. In other words, every one of their "actions" is part of a system and develops as systematically. That's why we consider them reliable. Therefore, reliability is a prerequisite for our trust in machines.

5 Machine Ethics

I wrote: as long as machines work, their procedural "actions" can be held to be reliable and can arouse an attitude of trust, of confidence, in us. When a device does not work, then we can consider the more complex system of which the device is part as reliable, so we trust such system. However, it might not work either. Not everything can be predicted and preordained. This is instead the belief that drives and guides many scientists in their investigations: the idea that, if not now, sooner or later everything will be preordained and therefore controllable. It's no coincidence that in 1948 Norbert Wiener gave the name of "Cybernetics"—from the name of the *kybernetes*, the helmsman who controls the route of a ship—to the discipline that was expected to turn such idea into reality (Wiener 2013). However, it does not only concern the ways in which scientific theories and technological practices are conceived. It also involves what the tradition of philosophical thinking calls "ethics".

In technological environments, it seems that one should only follow the right procedure to act well. In fact, as we have seen, this is what inspires that confidence and steadiness that makes us trust the "actions" of the machines. As a consequence, in technological environments, ethics—that is, the discipline that tries to govern the actions they must perform by interacting with an unpredictable world, based on principles shared by all rational subjects—is increasingly being replaced by a set of criteria that must be followed in a procedural manner. Therefore, codes of rules are replacing moral reflections. Abstract, general rules are mechanically applied to any particular circumstance. In other words: even in terms of actions, it seems that, if you want to act well, you only have to follow the procedure on the right terms.

Reflecting on the ethics of technological environments and its transformation into a rule of procedures is an important issue, if we want to gain an insight into trust. It is important, not only because, as we have seen, trust increases towards a regulated, predictable action. It is important, above all, because it makes us realise that there is another dimension that we have neglected so far, in our reflection. The question I asked at the beginning and that I am trying to answer—"can we trust machines?"—is not all about an analysis of the conditions that make one trust technological devices. Nor is it all about the psychological response—which can be a complex one, as it happens with the interweaving of trust and fear—that is triggered by some features of artificial agents. It also affects the way we consider, rate, judge the "action" of such agents. We can decide whether to trust such systems or not, and whether to trust them more or less, depending on how that happens.

As a matter of fact, like the actions of us humans, and the behaviour of our fellow creatures, the activities of machines too can be rated as "good" or "bad". That's why the experts speak of them precisely as of "artificial moral agents" (Wallach and Allen 2009). I'll say it again, this reference to the "morality" of machines is essential if we want to properly answer the question as to whether we can trust them or not. We actually believe, trust and rely only on what looks "good": on what or who, in other words, will not betray us, will meet its promises, is in tune with us, lets us achieve the purpose for which we have resorted to them. Only those we rate as good can actually be regarded as reliable too.

Of course, here we are far beyond the idea of procedural "goodness", in the sense of properly following a procedure. In what other senses can we then say that a machine is "good"? If we look at it per se, we can say that, insofar as values that we rate as "good" are embedded *in it*—that is, in the way it has been built and for the purposes it must accomplish (Flanagan et al. 2008). If we look at it *in terms of its operation,* we can say that, not only if it is "good" at performing the tasks it has been built and programmed for, but if such tasks can be held to be "good" too. In both cases, though, it's clear that an ethical judgement of the machine cannot be limited to the machine as such but should concern all the relations that make it what it is—those made by humans, about its design, its build, its programme, its maintenance—as well as all the consequences that such actions, also made by humans, have on the more or less autonomous activity of the machine. In a nutshell, machine ethics can only be conceived in relational terms: that is, considering the relationships, mainly with humans, which they are part of (Fabris 2016, 119–36).

We can get a better understanding of this dimension if we consider applying the ethical concept of "responsibility" to technological devices as well. If machines "act", then they are responsibility centres. But—again—in what sense? Are they directly or indirectly that? The answer is complex: both because a machine can independently "act" in varying degrees and because it can have many different interactions with humans.

It is humans, who have to do with machines in many different ways, who are responsible for them. Firstly, they are, because of those general principles by which the engineer, the builder and the programmer rule their work and which affect the choices that lead to the development of a specific piece of equipment. An example is a peace–loving engineer who won't certainly want to be involved in the development of combat drones. Secondly, they are, because of those criteria that motivate all such people to engineer, build and programme just that specific machine and that are therefore "embedded" in it, as we saw. Such criteria must factor in the scope and limits of the technological development of a given time, as well as the way such machine can be used or suggests to be used. Finally, in the relational perspective I am outlining here, humans are also responsible for the ways they connect an artificial agent with its environment and with the other beings in it, whether artificial or not: each of whom can act differently, according to different criteria and even in unpredictable ways.

Therefore, "responsibility" for the "actions" of a machine is always connected with responsibility for the actions of a human being. In this scenario, speaking of the "autonomy" of a machine and consequently of a "morality" of its "actions" takes a very specific meaning. We could say that a robot, for instance, can certainly *self–adjust* its processes, but cannot *self–regulate* them. In other words, it cannot "choose" the criteria and principles, according to which it relates with the environment, with other machines, with humans. It can only appropriate them. In other words, it can follow the criteria and principles according to which it has been built. It can also adjust its operations to some scenarios that it can anticipate. But, at least at the current stage of development, a robot cannot act on the principles according to which it has been built and works, or on the ways it interacts with its environment. This specific form of ethical self–reference is what is not happening yet, with a machine. It can correct its procedures but it cannot change the criteria such procedures have been built upon.

I'll say it again: for a machine, autonomy is about the way it can at most start its programmes, predict scenarios and respond to them. It is something that concerns its procedures and the ways they can be followed. It is therefore a "relative" autonomy:

relative to the criteria based on which the machine has been built, the specific context in which it works, the overview of predictable options, the ways its responses to specific environmental triggers are anticipated, and the rules that in some circumstances can be followed to achieve its goals. That's why we can say that a machine "acts" "well" or "badly" only in a limited sense. That's why we can trust it: in the sense of "trusting" it, not only for its procedural reliability, but for its embedded values and for the values of the human beings who have made it and who relate to it.

6 The Ethic Guidelines for Trustworthy Artificial Intelligence

In June 2018, the European Commission set up a High Level Expert Group (HLEG) to draw up *Ethic Guidelines for Trustworthy Artificial Intelligence* (see High-Level Expert Group on Artificial Intelligence 2019b). The first draft was presented in December that year and then, after a round of consultations, it was published in April 2019. After a trial period, during which the good practices contained in the third part of the text will be put to the test, a new version is expected to be issued in the first few months of 2020.

This document is extremely useful to discuss, test and substantiate the reflections I have developed so far. It can be an interesting case study for our purposes. It consists of 3 chapters. The first one lays down the general ethical principles and values that AI systems must adhere to. These are three principles, inspired by fundamental human rights: respect for human autonomy, prevention of harm, and fairness and explicability—i.e. transparency—of processes. The second chapter, which goes deeper into the details of the matter, lists another seven requirements that AI systems must meet to be considered reliable. They are: human supervision of such systems, technical reliability and safety, option to manage data and protect privacy, transparent processes, non-discrimination and fair use, maintaining environmental and societal wellbeing, potential accountability, meaning that such systems can be assessed and their impact managed. The last chapter then lists a number of specific points that should be used to find out, even more concretely, if such AI systems can be regarded as reliable or not. This is done by testing the criteria listed in the previous chapter.

The document is certainly more far–ranging than it looks from this brief summary. However, I cannot go into all the details. I would like, instead, to point out some of the basic assumptions and limits of its approach, partly in the light of my previous considerations.

The fact that trust in AI system is properly approached here, by setting it in the context of specific ethical principles, is definitely important. We can trust such devices, as we have seen, partly and especially because we judge them in ethical terms, not just in terms of performance. The problem is what idea of ethics the document refers to and what shared principles are invoked. We have seen them. They are those that, in the first chapter of the document, are associated with fundamental human rights and consist in respect for human autonomy, prevention of harm, and fairness and transparency of processes. Their importance is unquestionable. But what is not so clear in the document is whether they only concern the actions of human beings, such as engineers, builders, programmers, testers, users of technological devices, or if they also include the way such devices "act", in their relative autonomy. In other words, in the relationships between human agents and artificial agents, the role they play in building such relationships as bonds of trust under different circumstances is not clear.

We actually get the feeling that such problem can only be addressed and solved if humans take care of it, and more precisely if they are at the centre of such relationship. The web page that presents such document is aptly named *Building trust in human–centric AI*. But, can such approach be enough to dispel any fear of technological devices? To judge by that document, it seems the only way they can be trusted is when humans are back in control of them. But, as we have seen, this is not possible. AI systems are not merely at the service of humans: they are "artificial moral agents" themselves. Insisting on human control procedures might therefore be an inadequate solution, as it would not protect the trust of which the *High Level Expert Group*'s document would like to lay down the principles.

Then, there is another, even more crucial, point, which the document makes us reflect upon. What idea of a human being is it based on? And how is the relationship between such being and the AI devices perceived on the basis of such idea? The concept of "human" that is assumed there is, quite unsurprisingly, the same as the one developed in the modern age. It is a concept that puts the human being at the centre of every form of relationship, where every relationship is dependent on human will. In the text, as we have seen, such assumption is developed by reference to the fundamental human rights laid down by the EU. Yet, it is still an inadequate approach.

One cannot start from the idea that the human being is at the centre of all relationships with the beings that do not depend on him/her or with the systems he/she has built and programmed. Even more so because, I'll say it again, such systems have some degree of autonomy and so they are not always controllable. We can believe, of course, that their "actions" should be regarded as ethically good insofar as they are in the interest of humans—implicitly going back, in this case, to the first of Asimov's famous Three Laws of Robotics (Asimov 2004)—and the actions that AI devices perform when they go against such interests should be negatively judged. But, first and foremost, one should determine whether such interests concern all humankind or just a part of it. And then, more generally, one should keep in mind that the very idea of the human being as the centre of all his/her relationships had unacceptable, disastrous consequences. Think of the unrestrained exploitation of the whole ecosystem. That's why, rather than re–proposing a regulatory approach and listing a set of precepts like those listed in the EU document, one should completely rethink the underlying ethical perspective that should be taken on, in our relationships with artificial agents. That's the only way to justify our trusting them, as well.

7 Trust as the Quality of a Relationship

Now, we should answer our initial questions. Can we trust machines? In what forms and on what conditions can we trust them?

Let's recap the results of the previous considerations. We have found that nowadays we are led to trust AI machines, extending the notion of "trust" to our relationship with such devices as well. We have found how and why this happens. We have found that, in such circumstances, trust is always interwoven with a sort of fear. We have found that all this is linked to an ethical judgement that we can give of the "actions" of a machine and of the machine itself. Starting from there, we can give a better definition of what "machine ethics" does: it not only assesses the correctness of their procedures, it also considers the purposes they achieve through the relationships—with humans, with other machines—their "actions" are connected with. Finally, we have found that a recent attempt at working out some guidelines for a trustworthy AI—in a bid to manage the consequences of the actions that relate to it and the assumption of a specific idea of the human being—has a few limits.

All these reflections have brought to the fore at least two ideas that can help answer our questions, in the end. First, there is the fact that trust is an attitude that *only human beings* can have. We, not machines, are the only trustors. Machines can only be trusted, they cannot feel or display trust, albeit they might be designed to feign trust as some anthropomorphic robots do. Machines can only be trustworthy. But they are only within a very specific relationship.

This is actually the second important point that I want to dwell upon. Trust is born—as a human attitude towards someone or something that feels reliable—*only within a relationship.* Trust is a *specific quality* of such relationship. It is the way such relationship materialises.

What are the prerequisites for a bond of trust to be formed? They are those that see an entwining of similarities and differences in the terms of such relationship. I trust someone or something that is other than me, that can give me something I don't have, and that seems to be willing to give it to me, even if I can never be sure of this. That's why trust and fear are two sides of the same coin. Then, I trust someone or something insofar as they are somewhat similar to me and seem to be able to truthfully and faithfully respond to my openness. That's why, where openness and response meet, a bond of trust is formed and strengthened.

Such relationship can be rightly extended to technological devices as well, so, yes: *we can trust machines*. We can say we can trust them, not only because we are merely acknowledging what is happening nowadays, but also because machines as such deserve to be trusted. Extending the attitude that until now we had only held towards other human beings, some animals, a personal God, to such systems as well is legitimate, de facto and *de iure*. More than that. It is ethically legitimate.

Actually ethics, as we have seen, is the way we define our relationships based on criteria and principles that can be shared by all rational beings. Therefore, it is from an ethical perspective that the forms and conditions to give and govern trust can be established: even when we relate with artificial agents. Such forms and conditions, though, are not only established by humans, insofar as they try to retain control over the world. Such control is not always possible with devices that have an increasing level of autonomy. The forms and conditions I am speaking of are, then, those that are established by relational ethics: ethics that is focussed on the relationship between two or more terms, not just one of them, and that, in the ways such relationship is formed, finds the criteria to guide the actions of human beings and artificial systems alike, and to judge it. Only on this basis can a bond of trust be eventually created, and trust itself can prevail over fear.

References

Asimov I (2004) I Robot. Bantam Dell, New York

Botsman R (2018) Who can you trust? How technology brought us together and why it might drive us apart. PublicAffairs, New York

De Vaan M (2008) Etymological dictionary of latin and the other italic languages. Brill, Leiden-Boston

Fabris A (2016) Etica delle macchine, Teoria XXXVI/2016/2. Edizioni ETS, Pisa, pp 119–136

Fabris A (2018) Ethics of information and communication technologies. Springer, Berlin-New York

Flanagan M, Howe DC, Nissenbaum H (2008) Embodying values in technology. Theory and practice. In van den Hoven J, Weckert J (eds) (2018) Information technology and moral philosophy. Cambridge U.P., Cambridge-New York

High-Level Expert Group on Artificial Intelligence (2019a) A definition of AI: main capabilities and scientific disciplines. https://ec.europa.eu/digital-single-market/en/news/definition-artificial-intelligence-main-capabilities-and-scientific-disciplines

High-Level Expert Group on Artificial Intelligence (2019b) Ethic guidelines for trustworthy artificial intelligence. https://ec.europa.eu/futurium/en/ai-alliance-consultation/guidelines#Top

Hoad TF (ed) (2003) The concise Oxford dictionary of English etymology. Oxford U.P, Oxford-New York

Kroonen G (2013) Etymological dictionary of Proto-Germanic. Brill, Leiden-Boston

Lactantius (2004) Divine Institutes, translated by Bowen A Garnsey P. Liverpool U.P. Liverpool

Mori M (1970) The uncanny valley. Energy 7(4):33–35

Wallach W, Allen C (2009) Moral machines. Teaching robots right from wrong. Oxford U.P, Oxford-New York

Wiener N (2013) Cybernetics: or, control and communication in the animal and in the machine. Martino Fine Books, Eastford

Community and Trust in the Network Society. The Case of Virtual Communities

Veronica Neri

Abstract In the uncertain scenario that surrounds today's man, there is a new need for spaces to share, to establish new relationships of mutual trust. As the new technology advances, virtual communities are one of the responses to such need. In this revolutionary media environment, one might ask how a quite strong relationship of mutual trust can be created and nurtured in a virtual community, in relation to real objective relations and to the virtual community itself, as well as the technology that makes it possible, when it is, by definition, an ever-changing system. In this respect, some issues have been found to promote such relationships and others to (un)willingly hinder it. Therefore, the concept of trust should be re-semanticised to have a new relational paradigm built on ethical grounds, in which the individual becomes aware of what resources are required to cope with the insecurity of an increasingly uncertain society, partly with the help of virtual communities. Only through a jointly responsible behaviour, aimed at filling the gap that is inherent to the very concept of (virtual) community, can one trust something else—whether an individual, a technology or the community itself—aware of the hint of "magic" that is implied in the very etymology of the concept of trust.

Keywords Community · Virtual community · Ethics · Regulations · Trust

V. Neri (✉)
University of Pisa, Pisa, Italy
e-mail: veronica.neri@unipi.it

A. Fabris (ed.), *Trust*, Studies in Applied Philosophy, Epistemology and Rational Ethics 54, https://doi.org/10.1007/978-3-030-44018-3_10

1 Introduction

Over the last few decades, as an increasingly global and complex society was taking shape, individuals have felt cloaked in a climate of uncertainty that they try to appease in their own way. Concerns about economic instability, environmental risks and migration, as well as the problems of everyday life, fuel a culture of uncertainty and a fear of taking risks.[1] Typical of "reflexive modernity", such feeling cannot help affecting relationships between humans (Beck 1986; Beck et al. 1994; Fistetti 2003, 150). The individual choices that try to allay such concerns and affect everyone's private life can have fairly unpredictable consequences and can involve risks, but they can also unleash a wide variety of opportunities, regardless of our will or full awareness, for the life of other individuals, the community or society as a whole (Bauman 2007). Such climate, built on atomistic individualism and self-affirmation, has slowly undermined the role of the community (not just in a geographical sense) and the social connections that have been its distinctive trait throughout history. People live in a space, including the urban space, which consists of more superficial and fleeting relationships that aim to achieve short-term goals (Taylor 1989a, b, 2004). The sense of community, or the "desire for community" to use Bauman's words, then starts to reappear in response to the perception of this new fragility of the individual, in a media universe that has been revolutionised by the network, a *habitus* where messages can spread in a manner that is as quick as it is uncontrollable and sometimes even distorted (Bauman 2001; Fistetti 2003; Taylor 1991). It is a space that has dramatically changed the ways and models of relating with each other. And the virtual communities, which in such space have come to play a more and more important role—de-territorialised communities that have their own features and that, just like offline ones, are built on a relationship of mutual trust between their members—are a case in point (Rheingold 1993). But how can, then, the concept of trust be redefined in relation to the concept of community, today, in a global and "online" scenario that is more and more self-centred rather than "we-centred"? How can these new occurrences be reconciled with people's need to have more reassurances for themselves and for their life context, and thus build new trust (Luhmann 2002) in others (as individuals and as a community), in an uncertain, changing environment, where the members appear as profiles rather than as individuals? If one cannot rely on a far-reaching communal view (Taylor 1989a; MacIntyre 1981), an individualistic-narcissistic approach that isolates and withdraws individuals into themselves, deluding them into thinking they can be self-sufficient, but pushing them instead into self-effacement, seems to be unconceivable now. Or does an intermediate view of "holistic individualism", where the individual's roots in the community and therefore the relationship of trust

[1]From a historical-social point of view, uncertainty has actually always gone hand in hand with humans. What changes in today's world is that such phenomenon has reappeared at full strength in Western societies, after years of prosperity and financial security, such as the Seventies and Eighties (Millefiorini 2015, 288–291; Beck 1986; Sennett 1998; Giddens 1991; Inglehart 1998; Bauman 2001).

between the members of such community seem to be prerequisites for individuality to develop (Taylor 1989a, 154–155), apply to online communities as well, perhaps?

Virtual communities certainly do provide more freedom *from* the usual influences of objective reality to make room for self-accomplishment, but men are still responsible *for* their own actions and *for* others, meaning other individuals, a community and lastly a technology. Even if today's people aspire to *tout court* moral independence, it is only in a relationship and through a relationship and dialogue that they can construe themselves. Even a fluid community, a community built on the visible, as virtual communities are, where the winning rules are seeing/being seen and existing insofar as you are seen, in making all members play the same role, raises a need for a new bond of trust, which has to be readjusted to the new media landscape that sees technology becoming ever more independent. Maybe one should resort to that trust which, as in creed, comes from * kred, which brings to mind a sort of magical power with which we invest someone or something we trust. As opposed to objective reality, in the inter-reality of the network, even a conscious choice can turn into an unconscious one, against one's will (Natoli 2016, 40–41; Benveniste 1976, 131).[2]

2 Communities and Virtual Communities

Then, the meaning of the concepts of community and trust should be mentioned, first, to understand how their inextricable connections have changed since the advent of the network, therefore from a strictly media-centred perspective. Of course, a community, whether virtual or not, can only exist if it is essentially trusted and if its members trust each other, and they did even before it existed. To be perceived as a risk but also as an opportunity. More specifically, when speaking of online communities, one might ask whether they help encourage and/or strengthen relationships of trust among their members or not, and how this affects the objective reality.

---

[2]This paper was written with full and absolute awareness of the current trend towards a hybridization of our online and offline dimensions. The space within which we make our choices and within which we act or operate is increasingly permeated by the digital; however, from an ethical perspective, in order that we might foster the more positive aspects and, at the same time, curb those which are more negative, wemay go back to considering these as two separate dimensions, albeitmerely from a theoretical view point. Though the world of the so-called "virtual communities" is exceedingly vast and articulated, with various typologies which differ in terms of protocol and format, deontology and ethics can, however, help us to re-semanticize our relational dimension in these contexts. They can help us to understand more deeply and with greater awareness the reciprocity of influences between and with in such spaces. Hence, in this paper, we decided to start from one of the earliest definitions of the virtual community and to separate online and offline dimensions throughout the development of our discourse. This has enabled us to pin point certain behavioural aspects more typical of unconnected spaces as well as others more typical of digitally connected spaces, and indeed the possibilities they open up, or otherwise, in their ever-growing and inevitable fusion. This then leads us to how it can therefore be more 'virtuous' to make choices and to act within the current communicational context (Fabris 2018; Vitali Rosati 2012, pp. 152–153).

Going back briefly to the polysemic concept of community and the reasons that are reawakening interest in individuals, one cannot help noticing that, in philosophical terms, the network has coloured this word in new shades. Nowadays, individuals feel the social need to experience a sense of community that is up to date but does not affect their identity or independence.

From an etymological point of view, the word community comes from the Latin *communitas* and from the adjective *communis*, meaning something that is not one's own, that belongs to many, that is public as opposed to private. More specifically, we can separate the word into *cum*, a Latin preposition which conveys a social character, and *munus*, a noun, which also recalls the concept of duty. *Munus* brings to mind a gift where a return is eventually expected, a constant exchange that demands to be reciprocated (Esposito 2006, X–XII; Badellino and Caloghi 1989, 544). Actually, quoting Benveniste (1966), Fabris writes: "those "gifts" belonging to the same community are exchanged with confidence to consolidate their relationships. And among these gifts the word, or more generally the communicative act, is what most effectively achieves this purpose" (Fabris 2018, 10).

Such gift is therefore a sort of obligation that one undertakes towards someone else and where a discharge is expected. But nowadays individuals feel such only if they are released of such duty of *donum*, but this, however, wears away that relational and participatory dimension (as also recalled by the Greek *koinonia*, a noun associated with the concept of fellowship) (Esposito 2006, XVII–XVIII).

Oddly enough, despite this typical tendency of today's individualism, people are feeling again a longing for commonality, something that has always been our distinctive trait as relational beings and that, from the perspective of the history of thought, stems first and foremost from the political community of the Greek *polis* and from the *res publica* of the Roman tradition (Arendt 1958). The community still plays a key role in the Middle Ages, in the theological sphere, and is brought to an end, in the age of organicist communitarianism, by Ficino's theory and by the Protestant-inspired community of the Just. But it is the development of the modern State that brings to life a new way of looking at the community, as a political community or as an ethical community, for instance, through to the dialectic relationship between community and society, mainly studied by Tönnies (1887), Durkheim (1893) and Simmel (1890) at the turn of the XX century, and then that between community and association, studied by Weber (1922).

In more recent times, instead, apart from the diatribe between liberals and communitarians, a new sense of community is reawakened in the attempt to connect people with each other and with their identity. An "operative community" to be "with", once again, (Nancy 1983), because every human being attracts another human being or several (Blanchot 1983, 35).

But, with the new media, the individuals who live in the contemporary age need a new paradigm of community, suited to the overriding ambition to fulfil their wishes and goals, regardless and in spite of others. And, while confidence in the institutions is lost, trust is placed in people who are often unknown but who are "on an equal footing" within a "horizontally-organised" space (or at least seemingly so: Pasquale 2015). This brings to the fore a community of ordinary human beings (Agamben

2001), who can however actively participate in more or less well-organised virtual communities.

So, we wonder whether virtual communities may provide an answer to our need for immunising ourselves from life and from uncertainty, in a increasingly individualistic and self-centred society. And, so, as a consequence, we should try to find out how and to what extent the paradigm of mutual trust changes within and towards such new virtual set-up.

First and foremost, what does virtual community mean? According to the first definition provided in the literature and mentioned by Rheingold, it is a "social aggregation that emerges from the Net when enough people carry on public discussions long enough, with sufficient human feeling, to form webs of personal relationship in cyberspace" (Rheingold 1993, XX).

These set-ups are inhabited by individuals who share a common (deterritorialised) space and common goals, a feeling of belonging and an identity, and who, by interacting through the so-called new technology, build a bond of trust.

As opposed to the more traditional communities, here one can establish a synchronous interaction, as in chats and social networks, or an asynchronous one, as in forums or mailing lists. They are an alternative agent of socialisation which today's physical space seems unable to provide now. Freedom of expression seems wider and pushes back all those elements that can somehow engender (in)security (such as physical appearance, age, social class, etc.) in the traditional relational spaces of objective reality. All individuals can introduce themselves as they choose, replace their real identity with a different one, in a dialectical relationship that constantly reshapes and redefines itself or the relational space.

The member scan decide whether to be active or passive subjects. If they decide to be active, then they will send their own messages to introduce themselves or to share information, thus creating and promoting contents. If they decide to be passive, instead, they will only try to collect information and will want to see without being seen. These different states of visibility pave the way to a different relationship of trust, which affects objective reality, overlapping with and/or intersecting it, sometimes creating synergisms, sometimes conflict.

A space of seemingly greater freedom is created, in which some qualities of a subject can be valued or, by the same token, devalued.

While there are, then, mutual exchanges, trust in the other members, and a sense of belonging to and identification with the community (McMillan and Chavis 1986), in some extreme circumstances, such (unconscious) relatedness can slip into a false identity, where a profile other than the real one is presented. Moreover, some form of addiction may be experienced, with people preferring the dimension of virtual communities to that of objective reality, thus becoming more inept at establishing offline relationships. The new communal forms of the Net delude people into believing that they can control and handle such new phenomena in their own interest and that they can quit at any time. And, moreover, establishing a relationship of trust seems to be quicker and easier, though in fact it takes much longer, because, in feeling "protected" by a screen, people think they can have total control over the situation.

3 Trust and Virtual Communities

From a moral standpoint, what relationship can therefore be established between trust and a virtual community? It seems essential, therefore, to find out to what extent the relationship of mutual trust has changed inside and with regard to this new inter-reality, as well as between the members and the community, and in its interactions with objective reality. Trust, in which we are deeply rooted and that comes from our own childhood, remains a prerequisite that comes before trust itself (Natoli 2016, 16), as in the more traditional communities of the past, for such communities to form and, above all, to survive in the long term, profitably and beneficially (and not slip into misuse, such as cyber-bullying and hate speech). Trust "is a personal deed but even more than that it is a social fact, it shapes and governs people's relationships with their past– loyalty to inherited traditions as a *medium* for mutual acknowledgement – and with their future – meeting their obligations–; in other words, it ties memories and promises together". Only self-confidence can make one confide in someone else; but overconfidence can delude one into overestimating one's abilities and having problems confiding in others or deserving to be confided in Natoli (2016, 95). Actually, the degree of trust at stake here affects the person's choices and actions in the virtual community and offline.

So, what specific meaning can be attributed to trust, in this new virtual-community scenario, and what new relational models does it open up, within and without them?

Let's start with the etymological derivations of the word. The Romance root *fid*, as in the Latin *fides*, and the Greek *peith*, as in *peitho*, "to win over", "to believe", as well as "to persuade", both come from the Proto-Indo-European roots *bandh* (also in Sanskrit) and *bheidh*, which hint precisely to a social bond (Samek Lodovici 2019, 22; Romaniello 2004, 238; Rendich 2014, 271). Therefore, if one forms a relationship of trust, then one undertakes an obligation and is led to believe in what is communicated within such relationship of trust. And 'credere' (the Italian for believing) from the Indo-European * kred, as emphasised by Benveniste, is associated with a magical power (Benveniste 1976, 131). However, both the Latin *fido* and the Greek *peitho* also hint at the concept of persuasion, in both its active and passive forms, "as being persuaded" (*peithomai*). If we can arouse such feelings, then we can lead others to consider trusting us, and they need to trust or mistrust us before they can take any choice or action (Natoli 2016, 40–41).

In a virtual community, such trust is not only placed on the members or on the moderator, but also in the technology that makes it possible; even if we know that we do not trust them because we are perfectly aware of who the other really is or how the technology exactly moves—something that is increasingly getting out of our hand sand requires good digital literacy skills—but because of the possibilities that are inherent in such relationships, because we do not know the other well enough to have well-considered and perfectly informed expectations and because we have no way of knowing whether the underlying technological system will change or not, and how it will change over time.

The trust that we gained enables to form *online* relationships based on interconnections, "experienced from the point of view of mutuality" (Pastore 2014, 51 ff). In this way, we overstep that "natural trust" that we would place in your family and that, in the Net and especially in *social media*, we try to replicate in our connections between "friends" and "friends of friends". This fosters an "impersonal trust" that enables us to rely on other communities (and on their members)—instead of our own family or on other well-known circumstances that can feel us reassured—, or on the media system, though aware of the dangers and opportunities that could stem from some of our choices. Actually, if we are inclined to place more trust in what feels familiar, in the belief that if it worked before there are good chances that such choices will not fail in the future, we should go beyond that and embrace a conscious acceptance of the new risks, so as not to withdraw into indifference (Luhmann 2002, 27–33).

But we should not trust a virtual community, as opposed to a traditional community, just occasionally, when we are called to choose who to interact with or only the first time we come into contact with it, we should trust it every time we come into contact with it, and be aware and take account of all our actions and of the actions of the people we are interacting with. One could provide objective and verified information, but one could also provide misleading (if not downright false) information or information unsupported by safe sources and spread imaginaries and *modus vivendi* that can be influential by polarising thinking and, with it, behaviour. We can thus be led to believe in all the information that is closest to our imaginary and to our way of thinking unless we duly check the sources and look for a pluralist thinking that can only make us richer (Babad et al. 2012, 97–107). Otherwise, in a virtual community, one might enact relationships that do not embrace the freedom of informing or being informed, sharing and a feeling of belonging, but that spread around thoughts, goods and services that feel like brands as they force the members to conform (for instance by forcing them to give a pre-set number of replies or just a like or dislike).

Oddly enough, in such relational landscape it would be more appropriate, then, to weigh up one's choices more carefully. Therefore, trust should happen somewhere between an acquired awareness and the inevitable "unawareness" of some of the potential consequences of our actions, so that uncertainty must be necessarily be acknowledged. Such uncertainty is not necessarily negative, as it actually keeps us watchful and suggests a few ethical considerations by strongly recalling the concept of joint responsibility in a boundless communication community *à la* Apel (1992), a sort of communal citizenship that needs active participation to give support but also needs strength and a push to act for itself and for others, as part of a system that acts in real time. Otherwise, in a virtual community, one might betray the members' confidence and give rise to a phenomenon that is no longer *human oriented* and that only a conscious and shared responsibility can keep under control but that goes beyond that and (in)consciously "relies" on a sort of commodification of trust aimed to increase the number of users.

Of course, one may build very fragile relationships of trust, since users are shown more as profiles than as fully-fledged identities. Are were ally ready to respond to the other one, to take joint responsibility even for what we are not responsible for, to help human beings that we have never met before? The profile might go blank if the quality of information does not change over time, so that the profile take an identity of its own: in addition to information about social identities, we will eventually share information about our personal identity too, dressing it up in those nuances that distinguish it from the rest and make it feel more authentic, apart from the brief texts and above all the pictures. The higher the quality of the shared information, the more we prove to trust the other and the community, thus strengthening its essence (Fabris 2018, 32).

Of course, in a virtual community, trust get lost if the community is so wide it cannot be handled merely through trust; but we can opt for the purported reliability conveyed by the positive visibility that some communities have, in the Net. Their feedback acts as a sort of litmus test of conscious reliance, as we cannot rely on a definite identity or on someone who might not share the same cultural and social imaginaries.

4 Codes of Conduct, Between Netiquette and Deontology

As part of what we have outlined so far, trusting a virtual community is something that somehow disregards our actions, but that, nevertheless, must be put to the test every time we have access to a community. How can we lay down some criteria to give, earn and/or strengthen such feeling of confidence and to protect the members from each other and from the community, to steer clear of those ambiguous dimensions that the virtual space opens up?

Many virtual communities have moderators, for instance, or have drawn up their netiquettes, codes of conduct or codes of ethics that must be signed up and that state what fines are incurred for improper behaviour. But are such measures necessary and sufficient to impose a behaviour, *in* a community and *of* a community, that can be called virtuous and that will not deteriorate?[3] They can certainly help, but they are not enough to establish a relationship of mutual trust that makes one immune to potential failure and especially for the very essence of trust, which cannot be "bound" to trust.

As to the moderator (or moderators, depending on how big the virtual community is), when there is one, it acts as a supervisory authority: it can report inappropriate behaviour, delete messages or images that go against the purposes such community has been created for, based on specific algorithmic systems, on its own initiative or when reported by the users, so as to prevent the community slipping into hate speech, cyber-bullying, deep polarisation, etc.

In some circumstances, they opted for an etiquette (as well) (Rinaldi 1992; Fabris 2014, 126). It is a neologism that is a combination of the English "net" and the French "etiquette", meaning rules of conduct and habits that are dictated by experience and

[3]In this context, virtuous means respectful of the inspiring principles of an online community, such as the sense of belonging, sharing, mutual respect and truthful communication.

common sense and that must be adhered to within the community, to have consistently high-quality conversations and to avoid misunderstandings and conflict. Many versions of a netiquette have been drawn up (Livraghi 2001; Berlatsky 2013), all of them inspired by the 1855 RFC Guidelines 1995, laying down the rules for a good use of the network, and by the 2635 RFC Guidelines of 1999, as well as veritable codes of conduct. A code of conduct is a set of behavioural rules that must be enforced in an online community: it is drawn up and endorsed by the professional categories that write them and by the members of the community, and it is supported and laid down by the professional organisations that created the community, mostly in the form of prohibitions (Waller 2005). As Fabris writes "[T]he purpose of deontology is to establish the limits of informative and communicative activity through the indication of specific regulations and prohibitions. Not all, in fact, can be communicated. Precisely because this activity has a number of consequences, it is necessary to have clear what can be communicated and what does not. For example, we consider issues related to respect for privacy or not" (Fabris 2018, 15).

However, in legal terms, codes and netiquettes are worthless unless they are expressly mentioned in the Internet service providers' agreements. But all such measures, however essential to raise awareness of the need to behave respectfully within a community, seem to be fairly inadequate. There is no underlying reason for users to act according to specific rules and principles, which can only be based on the user's own sense of responsibility and therefore on ethics (Hart 1968; Raffoul 2010, 17–18; Vincent 2011, 15–35). A responsibility that can be defined as "a triangular relationship at the vertices of which are: firstly, *the responsible person* – the acting person", an independent individual or the member of a community who is free to act–; secondly, "*the sphere of responsibility*", in this case, the world of the virtual *community*; and, thirdly, "*the entity* or the person one has to answer to – the State, Parliament, individual conscience, a person or a community of people […]" (Franco 2015, pos. 55).

Despite such approach, the underlying technology and dynamics of a virtual community can still get out of hand. Fabris goes on to write: "These regulations are not totally in favour of the members of the community, who are simply offered the platform for interaction, but primarily it favours the interests of the companies that govern Social Networks. These interests are related to the acquisition of our information and data, and which we freely share while using a platform" (Fabris 2018, 76).

A bond of trust can still be broken against our will, but for merely procedural reasons that depend on the technology. But, most importantly, what is lacking in the deontological approach is a justification for the moral reasons on which our choices and then individual actions are based. The ethical issues raised by the virtual communities cannot be reduced to a strict compliance with behavioural rules and prescriptions (Fabris 2018, 15).

Without moral guidance, based on the responsibility that counterbalances individual freedom (increasingly tending, in the modern age, to fulfil one's own goals and wishes) and on meeting the obligation that typically ensues from a *donum*, there is no way of strengthening that bond of trust, which, in virtual communities, is the decisive factor, above and beyond a set of rules.

5 Conclusion

In the light of what has been discussed so far, however briefly, the relationship between virtual community and trust, because of the typical technological evolution of the so-called fourth revolution, may significantly and suddenly change, and therefore the members' options to choose and act can be affected accordingly.

So, how can a strong enough relationship of trust be maintained between individuals in a virtual community, in relation to objective reality and the virtual community itself, as well as the technology that makes it possible, when it is, by definition, an ever-changing system? Therefore, the concept of trust should be re-semanticised to have a new relational paradigm built on ethical grounds, in which the individual becomes aware of what resources are required to face the insecurity of an increasingly uncertain society, partly with the help of the virtual communities.

In this connection, some issues have been found to foster such relationship and others to (un)willingly hinder it.

Nowadays, these communities are built on visibility, within them, between the members, outside, in the general media community and in the offline world. Depending on the extent of its following, a virtual community can be felt as more exclusive and therefore more appealing. There is a proliferation of aesthetic virtual communities that follow the rules of the show business and that are built on a sort of emotional ataraxy, as well as the so-called "peg communities", where human beings temporarily 'lay down' their worries and receive 'collective' support. Such communities are as fluid as the societies they were born in, dealing with short-lived connections, whose members do not feel bound to any ethical responsibility, change their identity as the context changes, and tend to disappear as soon as the connection starts to play an important role in the users' emotional wellbeing (Bauman 2001, 70, 83). To maintain a strong relationship of trust, it should instead be put back at the centre of the relationship between oneself and the other, as an individual, as a community and as a technology, even by trying to find shared solutions to individual problems.

In this new virtual-community scenario, while the fragility of today's individuals cries out for trust, paradoxically, trust is more easily broken because online bonds can be broken more easily (despite problems deleting all the traces afterwards). In addition, if we think we can quit the community at any time without consequences, we might also have to deal with being totally discredited by our community, that could expose us, like an old-time scapegoat (Girard 1982, 1999), to digital pillory, which acts in unprecedented times and places, sometimes with tragic consequences for us, sometimes unknown to us. These communities are closed, made of "friends" or "friends of friends", or comparatively open, since they still need some acknowledgement, like those run by public institutions (Riva 2014, 108 ff). This comparative restriction enwraps users in a microcosm that is close to their interests. And, in this context, trusting a polarised virtual community can potentially be dangerous and can lead to extreme thoughts and behaviours.

So, based on what has been said so far, while there is some scepticism about online communities, Rheingold has no doubts that a sense of community is really experienced online, something that cannot but strengthen the relationship of trust. While trusting identities we do not know may be more complicated, our predisposition to interactions, with a view to achieving common purposes and a feeling of belonging to the same virtual community, built on a horizontal organisation in which everyone has the same chances to express themselves, can actually strengthen mutual trust. In this way, less formal processes are set off, and more significant information is shared, thereby strengthening the relationship of trust.

In addition, Rheingold says that: “There’s always another mind there. It’s like having the corner bar, complete with old buddies and delightful newcomers and new tools waiting to take home and fresh graffiti and letters, except instead of putting on my coat, shutting down the computer, and walking down to the corner, I just invoke my telecom program and there they are. It’s a place” (Rheingold 1993, 24). And this is only possible if there is a powerful motivation to take part in such new worlds, in the option to form impressions from the cues we find, and to understand the members’ feelings as conveyed through such channel. This is the only way not only to create weak contacts and bonds, but also to turn them into stronger and steadier bonds (McKenna and Green 2002; Riva 2014, 104–105).

The new technology, as Rivoltella writes, and Castell did before him, can no longer be considered as mere tools to communicate and share, but “at last they let themselves be perceived as a communal technology, insofar as they can form and maintain connections, veritable social synapses” (Rivoltella 2017, 8; Castells 2009).

To conclude, then, technology can certainly be an opportunity to re(constitute) the social capital of a community as well. Sharing and participation can be aimed at protecting mutual trust for the sake of communal trust, at breathing new life into the community as a whole, as well as strengthening trust between people, thus transcending the *solo-sphere* of the contemporary age (Dahlgren 2011; Dahlgren 2012, 11–12). However, we should not withdraw into those communities that oblige their members to meet requirements so unyielding that they will remarkably restrict their personal freedoms.

A desirable community would therefore be one in which the offline and the online world intersect, based on the *donum*, which involves mutuality, not just online but offline as well. The ultimate purpose of such participation (first and foremost, civil participation) should not be mere visibility as an end in itself but the strengthening of the relationship of trust.

The essential requirement for such process to take place in a virtual community is that we suspend our ethical responsibility. But, how can ethical communities be furthered where aesthetic communities prevail? An ethical community aspires to form long-lasting connections, with the necessary rights and duties, with responsible sharing and a commitment to respond to the members’ needs, not to create mere “gatherings of lonely souls” or slip into a virtual group (Bauman 2001, 66, 70; De Kerkhove 1995). The increasingly essential intersection between virtual communities and traditional communities engenders so much connectivity that mediation does not make any sense now. With the Internet of things (Eugeni 2015; Fabris 2018), virtual communities have joined a sort of nervous system *à la* Castells, where the media are the social "synapticspines" (Rivoltella 2017, 31). All the parts of this connective tissue are joined together to connect the different levels of the community, leaving the joints exposed. And the more an institution loses credibility in people's eyes, the more relevant such communities become. However, while such communities can foster a feeling of socialisation, they can messily shrink or expand, making mutual support more complicated (Maffessoli 1988, 28).

Therefore, within a virtual community, the right approach should be focussed on individual responsibility as well as on the joint responsibility of all its members, in the attempt to protect the community and not slip into indifference for oneself and for others, for the community and for technology in general. There is a risk that its quality standards may fall so low that the community will be totally closed. While it is true, then, that an absence of trust "would prevent one even getting up in the morning" (Luhmann 2002, 5), in such a virtual community, such trust is a "monitoring" and "monitored" trust, where we trust on a case-by-case basis, every time we log in, looking for constant confrontation—insofar as possible—with the communities of the objective world to join the two dimensions together and become more aware of our choices. As Luhmann writes: "where there is trust there are increased possibilities for experience and action, there is an increase in the complexity of the social system and also in the number of possibilities which can be reconciled with its structure, because trust is an effective form of complexity reduction" (Luhmann 2002, 11).

Trust must and can be put in a virtual community that ensures conscious relatedness, one that looks to oneself but that also looks to others, unconditionally, whoever they are, that makes it easier to recognise the members' identities by sharing information and services, but most importantly one that embraces supportive openness, not self-absorption and isolation, even if just for a limited time.

But this cannot be achieved without building a new paradigm on two dimensions: one that considers trust as a new social capital, the propeller of ethical growth, as it happens in the virtual communities of the public institutions; and one that calls the virtual space to enact that original sense of community, based on the dialectic relationship of mutuality that recalls precisely the etymological meaning of *munus*, "a giving that commits people to another in a joint obligation, and may we say, in an joint oath". That is the only way to maintain a strong bond of trust. The members of a virtual community cannot merely share an interest, they must share a liability, a gap to be filled, which nobody must be exempted from, in a potentially never-ending communicative loop (Esposito 2006, XII–XIII).

The keystone of ethical behaviour in a virtual community, if we are responsibly involved in it, as *cives*, is by keeping both dimensions in mind, to prevent the community from turning into an closed *habitus* and let item brace new perspectives, where showing our flaws may only be a spur to improve ourselves, to build constructive relationships.

Going back to the initial assumption, in this context, deep-rooted trust is more essential than ever, "trust comes before trust itself" (Natoli 2016, 16), but it goes hand in hand—and it cannot be otherwise—with risk, which can sometimes turn into an opportunity. We only need, therefore, to get back to that meaning of trust that comes from * kred and that calls a "magic force" to mind, that we bestow on someone we place trust on, whether it is another human being, the technological dimension or the communal dimension (Natoli 2016, 40–41; Benveniste 1976, 131).

References

Agamben G (2001) La comunità che viene. Bollati Boringhieri, Milano
Apel KO (1992) Etica della comunicazione. Jacabook, Milano
Arendt H (1958) The human condition. University of Chicago Press, Chicago
Babad E, Peer E, Hobbs R (2012) Media literacy and media bias: are media literacy students less susceptible to nonverbal judgment biases? Psychol Popul Media Culture 1(2):97–107
Badellino O, Calonghi F (1989) Vocabolario Latino-Italiano. Rosenberg & Sellier, Torino
Bauman Z (2001) Voglia di comunità. Laterza, Roma-Bari
Bauman Z (2007) Liquid times: living in an age of uncertainty. Polity Press, Cambridge
Beck U (1992) Risk society (1986). SAGE Publications Ltd., London
Beck U, Giddens A, Lash S (1994) Reflexive modernization: politics, tradition and aesthetics in the modern social order. Polity Press, Cambridge
Benveniste È (1966) Problèmes de linguistique générale. Gallimard, Paris
Benveniste È (1976) Il vocabolario delle istituzioni indoeuropee, I, Economia, parentela, società. Einaudi, Torino
Berlatsky N (2013) Netiquette and online ethics. Greenhaven Press, Detroit
Blanchot M (1983) La communauté inavouable. Les Éditions de Minuit, Paris, tr. it. by D. Gorret (2002) La comunità inconfessabile. SE, Milano
Castells M (2009) Communication power. Oxford University Press, Oxford
Dahlgren P (2011) The political web: media, participation and alternative democracy. Palgrave Macmillan, New York
Dahlgren P (2012) Social media and counter-democracy: the contingences of participation. In: Tambouris E, Macintosh A, Sæbø Ø (eds) Electronic participation. ePart 2012. Lecture notes in computer science, vol 7444. Springer, Berlin-Heidelberg, pp 1–12
De Kerkhove D (1995) The skin of culture investigating: the new electronic reality. Somerville House Publishing, Toronto
Durkheim È (1893) De la divisiondutravail social. Félix Alcan, Paris
Esposito R (2006) Communitas. Origine e destinodellacomunità, Einaudi, Torino
Eugeni R (2015) Semioticadei media. Le formedell'esperienza. Carocci, Roma
Fabris A (2014) Etica della comunicazione. Carocci, Roma
Fabris A (2018) Ethics of information and communication technologies. Springer International Publishing AG
Franco V (2015) Responsabilità. Figure e metamorfosi di un concetto, DonzelliEditore, Roma
Fistetti F (2003) Comunità. Il Mulino, Bologna
Giddens A (1991) The consequences of modernity. Stanford University Press, Stanford
Girard R (1982) Le boucémissaire. Grasset, Paris
Girard R (1999) Je vois Satan tombercommel'éclair. Grasset, Paris
Hart HLA (1968) Punishment and responsibility. Oxford University Press, Oxford
Inglehart R (1998) La società postmoderna. Mutamento, ideologie e valori in 43 paesi. EditoriRi-uniti, Roma
Livraghi G (2001) L'umanità dell'internet. Le vie della rete sono infinite. Tecniche Nuove, Milano
Luhmann N (2002) Vertrauen. Ein Mechanismus der Reduktion sozialer Komplexität (1968). Stuttgart, Lucius & Lucius, tr. it. by L. Burgazzoli (2002), La fiducia. Il Mulino, Bologna
MacIntyre A (1981) After virtue: a study in moral theory. University of Notre Dame Press, Notre Dame

Maffessoli M (1988) Le Tempsdes Tribus. Le déclin de l'individualisme dans les sociétés post modernes. Méridiens-Klincksieck, Paris, tr. it. by Zaramella A (1988) Il tempo delletribù. Armando, Roma
McKenna KYA, Green AS (2002) Virtual group dynamics. Group Dyn 6(1):116–127
McMillan DW, Chavis DM (1986) Sense of community: a definition and theory. J Community Psychol 14(1):6–23
Millefiorini A (2015) L'individuo fragile. Genesi e compimento del processo di individualizzazione in Occidente, Apogeo, Milano
Nancy JL (1983) La communauté désoeuvrée. Christian Bourgois, Paris
Natoli N (2016) Il rischio di fidarsi. Il Mulino. Bologna
Pasquale P (2015) The black box society: the secret algorithms that control money and information. Harvard University Press, Cambridge Mass
Pastore B (2014) Formedella fiducia nellospaziopubblico. In: Alici L, Pierosara S (eds) Generare fiducia. Franco Angeli, Milano, pp 55–72
Raffoul F (2010) The origins of responsibility. Indiana University Press, Bloomington and Indianapolis
Rendich F (2014) Dizionario etimologico comparato delle lingue classiche indoeuropee. Palombi, Roma
Rheingold H (1993) The virtual community. The MIT Press, Cambridge Massachusetts
Rinaldi AH (1992) The net user guidelines and netiquette. Computer User Services Florida Atlantic University, Sept 3. http://www.shentel.net/general/tiquette.html
Riva G (2014) Social network. Il Mulino, Bologna
Rivoltella PC (2017) Tecnologie di comunità. La Scuola, Brescia
Romaniello G (2004) Pensiero e linguaggio: grammaticauniversale. SoveraEdizioni, Roma
SamekLodovici G (2019) Fiducia e virtù. Teoria XXXVIII/ 2019/1, 117–136
Sennett R (1998) The corrosion of character, the personal consequences of work in the new capitalism. Norton, New York-London
Simmel G (1890) Über Sociale Differenzierung: Sociologische Und Psychologische Untersuchungen. Duncker & Humblot, Leipzig
Taylor C (1989a) Cross-purposes: the liberal-communitarian debate. In: Rosenblum NL (ed) Liberalism in the Moral Life. Cambridge University Press, Cambridge, pp 159–182
Taylor C (1989b) Sources of the self: the making of modern identity. Harvard University Press, Cambridge, MA
Taylor C (1991) The Malaise of modernity. Anansi, Toronto
Taylor C (2004) Modern social imaginaries. Durham, London
Tönnies F (1887) Gemeinschaft und Gesellschaft. Verlag di Fues, Leipzig
Vincent V (2011) A structured taxonomy of responsibility concepts. In: Vincent NA et al (eds) Moral responsibility. Beyond free will and determinism. Springer, Dordrecht, pp 15–35
Vitali Rosati M (2012) S'orienter dans le virtuel. Hermann, Paris
Waller BN (2005) Consider ethics: theory, readings, and contemporary issues. Pearson Longman, New York
Weber M (1922) Wirtschaft und Gesellschaft. Mohr, Tübingen

Trust and Political Institutions

Trust and Political Ignorance

Ilya Somin

Abstract We often hear claims that there is too much political distrust—so much so that society has become deeply divided and democracy itself may be imperiled as a result. At the same time, many also argue that citizens are too ready to believe "fake news," conspiracy theories, and other dubious material on political issues, thereby also potentially threatening the viability of democracy. These complaints are in some tension with each other. How can people be both lacking in trust, and highly gullible at the same time? But the tension is more apparent than real. The two problems are rooted in common dynamics of political ignorance and irrationality. Part I outlines the basic logic of rational political ignorance, and the related phenomenon of "rational irrationality"—biased evaluation of evidence. Part II applies this logic to current concerns about "fake news" and distrust of experts. In Part III, I explain why some substantial degree of deference to experts is desirable, and probably unavoidable. The key question is whether people can find the right information sources to trust. A number of strategies can help address this, but it seems unlikely many voters will adopt them without a change in incentives. Finally, Part IV outlines how we might have better incentives to assess facts and claims to expertise well if we make fewer decisions through ballot box voting, and more by "voting with our feet."

Keywords Democracy · Trust · Ignorance · Conspiracy · Fake news

---

Professor of Law, George Mason University. Parts of this paper are adapted from my forthcoming book, *Free to Move: Foot Voting, Migration and Political Freedom* (Oxford University Press), from Chap. 3 of my previous book, *Democracy and Political Ignorance: Why Smaller Government is Smarter* (Stanford: Stanford University Press, 2nd. ed. 2016), and from material previously printed at the Volokh Conspiracy law and politics blog, hosted by *Reason* magazine and the *Washington Post*.

I. Somin (✉)
George Mason University, 3301 Fairfax Dr., Arlington, VA 22201, USA
e-mail: isomin@gmu.edu

A. Fabris (ed.), *Trust*, Studies in Applied Philosophy, Epistemology and Rational Ethics 54, https://doi.org/10.1007/978-3-030-44018-3_11

1 Introduction

In the current political climate, we often hear claims that there is too much political distrust—so much so that society has become deeply divided and democracy itself may be imperiled as a result.[1] Some argue, also, that people more than ever distrust experts and others with specialized knowledge.[2]

At the same time, many also worry that citizens are too ready to believe "fake news," conspiracy theories, and other dubious material on political issues, thereby potentially threatening the viability of democracy in a different way.[3] These complaints are in some tension with each other. How can people be both lacking in trust, and highly gullible at the same time?

But the tension is more apparent than real. In this article, I explain how the two problems are rooted in common dynamics of political ignorance and irrationality in the evaluation of information. In a democracy, most voters have little incentive to either acquire more than minimal political knowledge, or objectively evaluate the information they do learn. Such ignorance and illogic is actually rational behavior from the standpoint of individual voters. But it has harmful systemic effects.

The upshot is that, thanks in large part to ignorance and illogical thinking, voters often place their trust in the wrong places, while ignoring higher-quality information sources. The problem is not so much that there is too little or too much trust, overall, as that trust is often misplaced.

On many issues, voters necessarily lack expertise, and therefore must to some extent defer to the views of those more knowledgeable than themselves. Trust in such information sources is neither good nor bad, in and of itself. It all depends on how worthy they are of that trust.

The key question is whether voters can find the right information sources to trust, without falling prey to deceptions, distortions, "fake news," and bogus claims of expertise. There is much that individual voters can do to improve their performance on this score. But I am not optimistic that it will actually happen absent systemic change in incentives.

Part I of this article outlines the basic logic of rational political ignorance, and the related phenomenon of "rational irrationality"—biased evaluation of the evidence voters do learn. I also explain why the combination of these factors can both undermine political trust, and also lead voters to be gullible about accepting a variety of dubious claims that reinforce their preexisting views.

---

[1] See, e.g., Liliana Mason, *Uncivil Agreement: How Politics Becomes Our Identity*, (Chicago: University of Chicago Press, 2018); Darrell M. West, *Divided Politics, Divided Nation: Hyperconflict in the Trump Era*, (Washington, DC: Brookings Institution, 2019).

[2] See, e.g., Tom R. Nichols, *The Death of Expertise: The Campaign Against Established Knowledge and Why it Matters*, (New York: Oxford University Press, 2017).

[3] See, e.g., Russell Muirhead and Nancy Rosenblum, *A Lot of People are Saying: The New Conspiracism and the Assault on Democracy*; James Ball, *Post-Truth: How Bullshit Conquered the World*, (New York; Biteback, 2018): Cailyn O'Connor and James Owen Weatherall, *The Misinformation Age: How False Beliefs Spread*, (New Haven: Yale University Press, 2018).

Part II applies this logic to current concerns about "fake news" and distrust of experts. It shows how rational ignorance and irrationality lead us to distrust information that cuts against our preexisting views, while being receptive to even very dubious claims that reinforce them, including conspiracy theories. This may be exacerbated by the internet and other new technologies, but it is far from unique to them.

In Part III, I explain why some substantial degree of deference to experts is desirable, and probably unavoidable. I also discuss ways in which voters can potentially do a better job of assessing supposedly expert information sources. It would help if they were careful about evaluating the scope of the supposed experts' actual knowledge, if they tried to correct for potential ideological bias, and if they separated out factual issues (on which experts are likely to have superior insight) from purely normative ones (on which they often will not). But I also offer some reasons why it is unlikely that very many voters will actually improve their judgment of such issues.

Finally, Part IV briefly outlines how we might have better incentives to assess facts and claims to expertise well if we make fewer decisions through the political process, and more by "voting with our feet." When people vote with their feet in the private sector, or by deciding what jurisdiction to live in within a federal system, they have much stronger incentives to both seek out relevant information and evaluate it objectively.

2 Rational Ignorance and Rational Irrationality

If your only reason for following politics is to be a better voter, that turns out to not be much of an incentive at all, because there is so little chance that your vote will actually make a difference to the outcome of an election—about 1 in 60 million in a US presidential race, for example.[4] For most people, therefore, it is rational to devote very little time to learning about politics, and instead focus on other activities that are more interesting, or more useful. As former British Prime Minister Tony Blair put it, "[t]he single hardest thing for a practising politician to understand is that most people, most of the time, don't give politics a first thought all day long. Or if they do, it is with a sigh…., before going back to worrying about the kids, the parents, the mortgage, the boss, their friends, their weight, their health, sex and rock 'n' roll."[5] Most people do not precisely calculate the odds that their vote will make a difference. But they can develop an intuitive sense that the likelihood is small, and

[4] Andrew Gelman, Nate Silver, and Aaron Edlin, "What is the Probability that Your Vote Will Make a Difference?" *Economic Inquiry* 50 (2012): 321–26.

[5] Tony Blair, *A Journey: My Political Life*, (New York: Alfred A. Knopf, 2010), 70–71.

act accordingly. They have an incentive to be "rationally ignorant"—to devote little or no time and effort to learning about government and public policy.[6]

Survey data shows that rational ignorance is an accurate characterization of most voters' level of political knowledge. They often lack even very basic knowledge about the candidates and policy questions at issue in any given election.[7]

For much the same reasons as they have incentives to be rationally ignorant, most voters also often have little or no incentive to analyze the information they do learn in a logical, unbiased way. To the contrary, voters have incentives to fall prey to "rational irrationality": when there are few or no negative consequences to error, it is rational to make almost no effort to control one's biases.[8] For example, voters routinely overvalue any evidence that supports their preexisting views, while downplaying or ignoring anything that cuts the other way.[9]

An additional extension of this concept is the problem of "motivated ignorance," where people choose to avoid certain information not because it is too costly to acquire, but because there are psychic costs associated with knowing it.[10] Knowing a "painful" truth might make person in question more knowledgeable and better able to analyze issues. But these benefits may not be worth the psychic pain of calling into question treasured beliefs, undermining a sense of identity, or perhaps causing conflict with members of a given peer group. Thus, it may be rational to avoid such information in cases where the costs of knowledge are high relative to potential benefits. This can occur in a wide range of situations, but is especially likely with political information, where the benefits of improving knowledge—for most voters—are extremely low.[11] Thus, voters may avoid acquiring knowledge that is likely to challenge their preconceived notions or to create conflict between them and members of their "identity groups," including racial, ethnic, and religious groups, and partisan supporters of the same political party.[12]

In this paper, I treat "motivated ignorance" as a special subset of rational irrationality. Both are situations where it is instrumentally rational for a person to do a poor job of evaluating the epistemic significance of information—either information they already know or information they could potentially acquire. But it can also

[6]For a more extensive discussion of the logic of rational ignorance, see Ilya Somin, "Rational Ignorance." in *Routledge International Handbook of Ignorance Studies*, (Matthias Gross and Linsey J. McGoey, eds., Routledge, 2015). The idea was first developed by Anthony Downs, *An Economic Theory of Democracy*, (New York: Harper & Row, 1957).

[7]This part of the article builds on my book *Democracy and Political Ignorance: Why Smaller Government is Smarter*, (Stanford: Stanford University Press, 2nd ed. 2016), which analyzes rational ignorance and its consequences in great detail (see esp. Chaps. 1–4).

[8]See ibid., Chap. 3, and Bryan Caplan, *The Myth of the Rational Voter*, (Princeton: Princeton University Press, 2007).

[9]For a review of the evidence, see Somin, *Democracy and Political Ignorance*, 92–97.

[10]See Daniel Williams, "Motivated Ignorance, Rationality, and Democratic Politics," Synthese (forthcoming, 2020), available at https://link.springer.com/content/pdf/10.1007%2Fs11229-020-02549-8.pdf.

[11]Ibid, 12–17.

[12]Ibid.

be seen as a distinct phenomenon.[13] Either way, it creates additional incentives for voters to make epistemic mistakes.

Rational ignorance and rational irrationality affect the decisions of altruistic voters, as well as those who are narrowly self-interested. Even a citizen who is strongly motivated to help others still has little incentive to devote more than a small amount of effort to acquiring political knowledge and trying to rein in his or her biases. Whether her purposes are self-interested or not, the odds that her efforts will pay off is extremely low. This makes it rational for both egoists and altruists to severely limit the time and effort devoted to acquiring and analyzing political information.[14]

Rational ignorance does not necessarily require careful, calculated decision-making. In many cases, it involves merely application of crude rules of thumb or an intuitive sense that there is little benefit to seeking out additional knowledge. Thus, the idea is not dependent on the assumption that voters are hyperlogical or capable of making complex calculations about odds. Indeed, such detailed calculation may itself be irrational, since it may require more time and effort than can be justified given the likely benefit.[15]

Decades of survey data indicate that voter knowledge levels are low, and have experienced little or no increase despite rising educational attainment, and the development of the internet and other modern technology that makes information easier to access.[16] Often, the majority of the public does not know even basic information, such as which party controls Congress, what major policies have been enacted, or which elected officials are responsible for which issues.[17] Just before the 2014 election, in which the main stake at issue was control of Congress, only 38% of voters knew which party controlled the House of Representatives, and a similar percentage knew which controlled the Senate.[18] Another 2014 survey found that only 36% of Americans can even name the three branches of the federal government: the executive, the legislative, and the judicial.[19] While political ignorance has been more

[13]For an argument for its distinctiveness, see ibid., 4–10.

[14]For a more detailed discussion, see Somin, *Democracy and Political Ignorance*, 78.

[15]For more detailed discussion of and the reasons why rational ignorance does not require careful calculation and is consistent with crude heuristics, see Ilya Somin, "Rational Ignorance." in *Routledge International Handbook of Ignorance Studies*, (Matthias Gross and Linsey J. McGoey, eds., Routledge, 2015); and Brad R. Taylor, "The Psychological Foundations of Rational Ignorance: Biased Heuristics and Decision Costs," University of Queensland, unpublished paper, Aug. 31, 2019, available at https://papers.ssrn.com/sol3/papers.cfm?abstract_id=3443280.

[16]For recent overviews of the evidence, see, e.g., Somin, *Democracy and Political Ignorance*, Chap. 1; Jason Brennan, *Against Democracy*, (Princeton: Princeton University Press, 2016); Christopher Achen and Larry Bartels, *Democracy for Realists: Why Elections do Not Produce Responsive Government*, (Princeton: Princeton University Press, 2016); Rick Shenkman, *Just How Stupid Are We? Facing the Truth About the American Voter*, (New York: Basic Books, 2008).

[17]For numerous examples, see Somin, *Democracy and Political Ignorance*, Chap. 1.

[18]Ibid., 1.

[19]Ibid., 20. Inmore recent surveys, the number who can name the three branches varies from 26 to 39%. Annenberg Public Policy Center survey, Aug. 9–13, 2017, available at https://www.annenbergpublicpolicycenter.org/americans-are-poorly-informed-about-basic-constitutional-provisions/ (26%); Annenberg Public Policy Centersurvey, Aug. 16–27, 2019 (39%),

thoroughly studied in the USthan in other countries, survey data indicates that there is similar ignorance in other democracies. It is not a problem unique to the United States.[20]

An informed electorate is a public good, in the economic sense of the term[21]: people benefit from its production even if they have not contributed to its creation, each individual's contribution is infinitesimally small, and the benefits are "nonrivalrous"—my enjoyment of them is not reduced by that of other members of society and vice versa. Like many other public goods, it is tends to be underproduced, because individuals have strong incentives to underinvest in it.

Not all political ignorance is rational: some may be inadvertent—a case of voters failing to even consider the possibility that there might be relevant information they don't know that could help them improve the quality of their decisions.[22] But rational ignorance theory best explains why such a high percentage of the voting public is routinely unaware of even very basic facts about government and public policy (whose potential usefulness is hard to overlook), and why political knowledge levels have not risen in recent decades despite the greater availability of information, which should make its utility harder to overlook.[23]

Some scholars argue that political ignorance is not a significant problem, because voters can offset their lack of knowledge by relying on information shortcuts, or because electorates made up of poorly informed individual voters still make good decisions in the aggregate.[24] I cannot address these issues in detail here. But I have criticized both shortcut theories and "miracle of aggregation" arguments in considerable detail in my book, *Democracy and Political Ignorance*.[25]

---

available at, https://www.annenbergpublicpolicycenter.org/americans-civics-knowledge-increases-2019-survey/

[20]See, e.g., Ipsos-MORI, *Perils of Perception: A Fourteen- Country Study* (Ipsos-MORI, 2014) (detailing similar ignorance in many leading democracies); Bobby Duffy, *The Perils of Perception: Why We're Wrong About Nearly Everything*, (London: Atlantic Books, 2018), Chaps. 4–9.

[21]For more detailed discussion, see ibid., Chap. 3.

[22]For a recent defense of this view of political ignorance, see Jeffrey Friedman, *Power Without Knowledge: A Critique of Technocracy*, (New York: Oxford University Press, 2019). For a more detailed discussion of the difference between rational ignorance and inadvertent ignorance, see Somin, "Rational Ignorance," 275.

[23]For more extensive discussion of these points, see Somin, *Democracy and Political Ignorance*, 89–91.

[24]See, e.g., Donald Wittman, *The Myth of Democratic Failure*, (Chicago: University of Chicago Press, 1995); Hélène Landemore, *Democratic Reason: Politics, Collective Intelligence, and the Rule of the Many*, (Princeton: Princeton University Press, 2013); James Stimson, "A Macro Theory of Information Flow." in John Ferejohn and James Kuklinski, eds., *Information and Democratic Processes*, (Urbana: University of Illinois Press, 1990); Bernard Grofman and Julie Withers, "Information-Pooling Models of Electoral Politics." in Bernard Grofman, ed., *Information, Participation and Choice*, (Ann Arbor: University of Michigan Press, 1993); James Surowiecki, *The Wisdom of Crowds: Why the Many Are Smarter than the Few*, (New York: Doubleday, 2004), Chap. 12; and Robert S. Erikson, Michael B. Mackuen, and James A. Stimson, *The Macro Polity*, (New York: Cambridge University Press, 2002).

[25]Somin, *Democracy and Political Ignorance*, Chap. 4.

3 Ignorance, Distrust, and Gullibility

The combination of rational ignorance and rational irrationality make citizens distrust experts and others who convey information that undermines their preexisting views. But they also make them more susceptible to political misinformation and deception that reinforces those predispositions.

The distrust is not difficult to explain: thanks to rational irrationality, voters dislike and distrust any information—including claims by supposed experts—that cut against their preexisting views. Distaste for information that challenges one's preexisting views is a natural human emotion. And in the world of political information, there is little incentive to keep such reactions in check, because there is little cost to the individual voter for getting things wrong. Motivated ignorance exacerbates the problem, as it creates additional incentives to avoid information sources that might challenge one's preconceptions.

Studies repeatedly show that voters are susceptible to misinformation that reinforces their preexisting views, but often resistant to correction of their factual errors.[26] A more optimistic recent analysis finds that voters are often willing to accept factual corrections of misstatements and "fake news" stories presented in controlled experiments.[27] But most voters do not in fact systematically seek out corrections to their factual beliefs in the way that researchers can do with their experimental test subjects. To the contrary, they tend to select information sources that reinforce rather than challenge their preexisting views.[28]

Voters' tendencies towards biased evaluation of information are exacerbated in periods of severe and increasing ideological and partisan polarization,—like that which we are currently experiencing in the United States and some European nations.[29] In such situations, facts and arguments presented by partisan adversaries are viewed with even greater hostility than normally, because they are seen as emanating from bad people, not just those with disagreements. In addition, they are likely to pose more fundamental challenges to our worldviews than criticisms or corrections emanating from opponents who are perceived as less distant from our own worldview.

These consequences of rational irrationality are exacerbated by pure rational ignorance, as well. A voter who is highly knowledgeable about a given set of issues may be able to find merit in an opposing view, because her preexisting understanding might lead her to recognize that the opponent has made a valid point, even despite

[26]See, e.g., ibid., 94–97 (summarizing evidence); Charles S. Taber and Milton R. Lodge, "Motivated Skepticism in the Evaluation of Political Beliefs," American Journal of Political Science 50 (2006): 755–69; Brendan Nyhan and Jason Reifler, "When Corrections Fail: The Persistence of Political Misperceptions," Political Behavior 32 (2010): 303–30.

[27]Ethan Porter and Thomas J. Wood, *False Alarm: The Truth About Political Mistruths in the Trump Era*, (New York: Cambridge University Press, 2019).

[28]For an overview of evidence on this point, see Somin, *Democracy and Political Ignorance*, 95–96.

[29]See, e.g., Mason, *Uncivil Agreement*.

the temptation to reject it. This is less likely if she has little background knowledge to begin with, and therefore cannot readily tell strong points from weak ones.[30]

The same forces that make voters distrust claims put forward by political adversaries also make them more gullible with respect to dubious and even outright bogus claims made by their allies. Ignorance makes it difficult to accurately assess the validity of such claims. And rational irrationality creates a strong temptation to accept them if they reinforce our preexisting views.

Ignorant voters may believe outlandish claims that they would reject if they had the benefit of greater knowledge. Similarly, a person who analyzes political information in a highly biased way is more likely to accept falsehoods that reinforce his preconceived views.

A variety of conspiracy theories and other highly dubious misinformation is believed by large portions of the American public. The election of Trump in 2016, and his penchant for conspiracy theories, highlighted this problem.[31] But the trend long predates Trump and is unlikely to end with his departure from the political scene.[32]

A survey conducted by economist Neil Malhotra and political scientist Yotam Margalit in early 2009 showed that some 25% of non-Jewish Americans believe that "the Jews" deserve at least "a moderate amount" or "a great deal" of blame for the financial crisis of 2008.[33] Some 32% of self-identified Democrats and 18% of Republicans endorsed that view.[34] A March 2010 Harris Poll survey found that 25% of Americans endorse the "birther" claim that President Barack Obama "was not born in the United States and so is not eligible to be president."[35] About 45% of self-described Republicans, 24% of independents, and 8% of Democrats endorsed birtherism. Donald Trump was a prominent promoter of birtherism at the time[36]; but it would be a mistake to ascribe its popularity solely or even primarily to his advocacy.

One can find comparable examples of misinformation on the political left. A 2007 poll of likely voters found that 35% of Democrats believed that President George

---

30 I explore this point in Somin, *Democracy and Political Ignorance*, 99–104.

31 See, e.g., Muirhead and Rosenblum, *Many People are Saying*; Peter Nichols, "Trump Needs Conspiracy Theories," *The Atlantic*, Nov. 29, 2019, available at https://www.theatlantic.com/politics/archive/2019/11/trump-conspiracy-theories-ukraine/602728/.

32 For a historical overview of conspiracy theories in American politics see Joseph Uscinski, *American Conspiracy Theories*, (New York: Oxford University Press, 2014).

33 Neil Malhotra and Yotam Margalit, "Anti-Semitism https://www.theatlantic.com/politics/archive/2019/11/trump-conspiracy-theories-ukraine/602728/. And the Economic Crisis," *Boston Review*, May–June 2009, available at https://bostonreview.net/BR34.3/malhotra_margalit.php.

34 Ibid.

35 Harris Poll survey, March 1–8, 2010, available at https://www.harrisinteractive.com/vault/Harris-Interactive-Poll-Research-Politics-Wingnuts-2010-03.pdf.

36 For an overview, see Andrew Prokop, "Trump Fanned a Conspiracy Theory About Obama's Birthplace for Years. Now He Blames Hillary Clinton for it," *Vox*, Sept. 16, 2016, available at https://www.vox.com/2016/9/16/12938066/donald-trump-obama-birth-certificate-birther.

W. Bush "kn[e]w about the 9/11 attacks in advance," while 26% were not sure.[37] By comparison, only about 12% of Republicans and 18% of independents took the same view.[38]

Belief in many other political conspiracy theories is common as well, including claims that the government is hiding evidence of visitation by alien civilizations, claims that the AIDS virus was deliberately manufactured to target African-Americans, and assertions that government agencies planned the assassinations of President John F. Kennedy and other prominent political leaders.[39]

Rational ignorance is surely a contributing factor to belief in this kind of misinformation. Political knowledge tends to be highly correlated with education, and less educated survey respondents were more likely to endorse birtherism, 9/11 conspiracy theories, and the claim that "the Jews" are to blame for the financial crisis.[40] People who are unfamiliar with the normal functioning of the political system and economy are more likely to find dubious disinformation plausible than those with greater knowledge.

For example, one of the factors that makes "birtherism" implausible to more knowledgeable voters is that, if Barack Obama were genuinely constitutionally ineligible to become president, his political opponents would have had strong incentives to uncover this fact and use it to force him to withdraw from the race. Both Obama's Democratic primary opponent Hillary Clinton and his Republican general election opponent, John McCain, had highly professional, well-funded campaigns. If there was any real evidence proving that Obama was not born in the United States, they would likely have uncovered it. Reasonably well-informed voters would have understood this and discounted "birtherism" accordingly.

9/11 conspiracy theories are implausible for similar reasons. For the Bush Administration to have deliberately chosen to let the attacks happen despite knowing about them in advance, the president would have had to organize a conspiracy involving numerous officials in the White House, intelligence agencies, the Department of Defense, the armed forces, and elsewhere. Given the highly leak-prone nature of American government, the overwhelming likelihood is that any such conspiracy would be leaked, especially given its morally outrageous nature. Some official in the know would surely have revealed the information, whether for moral reasons or a desire to avoid criminal prosecution for complicity in the plot if its existence were later were revealed by others.

And the revelation that high-ranking officials deliberately chose to allow a terrorist attack that killed thousands of Americans would have destroyed the Administration

---

[37] Rassmussen poll, April 30-May 1, 2007.

[38] Ibid.

[39] Each of these examples is noted in Cass R. Sunstein and Adrian Vermeule, "Conspiracy Theories," University of Chicago Law School Law & Economics Research Paper No. 387 (2008); See also Sunstein, *Conspiracy Theories and Other Dangerous Ideas*, (New York: Simon and Schuster, 2016).

[40] Malhotra and Margalit, "Anti-Semitism and the Financial Crisis"; Rassmussen poll, April 30-May 1, 2007; Harris Poll survey, March 1–8, 2010, available at https://www.harrisinteractive.com/vault/Harris-Interactive-Poll-Research-Politics-Wingnuts-2010-03.pdf.

politically, and severely damaged the Republican Party as a whole. Even if the President and his advisers were sufficiently amoral to consider such a conspiracy, they would quickly realize that the political risks far outweighed any possible gain.

Finally, the claim that "the Jews" caused the financial crisis is implausible to objective observers with even a passing knowledge of the standard economic explanations for the crash. Moreover, it is not clear why Jews would want to cause a financial collapse in which Jewish investors and financiers suffered no less than gentiles.[41]

In all these cases, basic political and economic knowledge could greatly reduce the likelihood of believing such implausible falsehoods. Rationally ignorant voters who lack such knowledge, however, are more susceptible to believing them.

Rational irrationality also deserves some of the responsibility. It is probably no accident that Republicans are disproportionately susceptible to birtherism, while Democrats are far more likely to endorse 9/11 conspiracy theories. Obviously, it is no secret that partisan Republicans tend to be hostile to Obama, while most partisan Democrats felt similarly about Bush. These predispositions make partisans more willing to believe any claim that reflects poorly on their political enemies—often without carefully considering whether the claim is true or even plausible.

Such bias seems irrational if the partisans' goal is to get at the truth, to determine whether the allegations against Bush or Obama are accurate. But it is perfectly rational if their objective is at least partly to enjoy the emotional satisfaction of being confirmed in their preexisting views.

4 When Should Voters Defer to Experts?

One possible alternative to rational irrationality is simple across-the-board skepticism about all claims about political issues. Perhaps rationally ignorant voters should just presumptively reject any assertions about public policy, except those few on which they happen to develop extensive knowledge. They should not give any special credence to real or imagined experts, but should only trust either first-hand knowledge, or data that they have had the opportunity to carefully assess for themselves.

But such a stance not only requires great self-discipline to avoid forming opinions, it is also unrealistic. The world is so complex, and there is so much information out there, that deference to experts on many issues is unavoidable. In many aspects of life, we inevitably have to make decisions based on deference to the superior knowledge of others. For example, few do their own scientific research on the effectiveness of various medicines or vaccines. We instead defer to medical experts, such as doctors

[41] A possible alternative scenario is that Jews deserve "blame" for the crisis because Jewish financiers or government officials caused it inadvertently, rather than deliberately. However, there is no evidence to suggest that Jewish officials and financiers behaved differently from gentile ones in this respect.

and scientists. The same goes for many other decisions we make every day. In a world of specialization and division of labor, it would be irrational to try to develop our own expertise on every issue on which we have to make decisions.[42]

For similar reasons, it is unrealistic to expect voters to achieve expert-like understanding of most political issues in their own right. There are far too many political issues out there, and few if any citizens have the time, resources, and cognitive ability needed to become highly knowledgeable about more than a small fraction of them.

Rational ignorance combined with a degree of deference to expertise is a necessity. The question is how to figure out which experts to trust, and which ones to be suspicious of. A 2015 Pew Research Center study showing disagreement between scientists and the general public provides a helpful basis for considering this question.

The survey shows that scientists and the general public disagree on a wide range of science-related public policy issues.[43] For example, it finds that 87% of scientists, but only 37% of the general public believe that it is safe to eat genetically modified foods; 68% of scientists believe it is safe to eat food treated with pesticides, compared to only 28% of the public.[44] Relative to the public, scientists are much more supportive of nuclear power and the use of animals in scientific research, and much less supportive of offshore drilling. Also, some 87% of scientists believe that climate change is mostly due to human activity, a view shared by only 50% of the public.[45]

A. The Case for Deferring to Scientists

This raises the question of whether voters should defer to majority scientific opinion on these issues. Given evidence of widespread ignorance, it is tempting for me to conclude that the answer is almost always "yes."[46] The majority of the public is often ignorant about basic facts about government and politics, and their scientific knowledge is also far from impressive.[47] You do not have to believe that scientists are always right about scientific issues to conclude that they are on average *more likely* to be right than generally ignorant voters are. To the extent that this is true, an electorate that defers to majority scientific opinion on these issues would make fewer mistakes than one that does not, even though neither would be completely error free.

The above reasoning has some merit. But it is important to avoid conflating two different kinds of "scientific" issues. Some of the questions addressed in the Pew

---

[42]I discuss these points in greater detail in Ilya Somin,, "Rational Ignorance." in *Routledge International Handbook of Ignorance Studies*, (Matthias Gross and Linsey J. McGoey, eds., Routledge, 2015); cf. Brad R. Taylor, "The Psychological Foundations of Rational Ignorance: Biased Heuristics and Decision Costs," University of Queensland, unpublished paper, Aug. 31, 2019, available at https://papers.ssrn.com/sol3/papers.cfm?abstract_id=3443280.

[43]Pew Research CenterSurvey, Jan. 29, 2015, available at https://www.pewresearch.org/science/2015/01/29/public-and-scientists-views-on-science-and-society/.

[44]Ibid.

[45]Ibid.

[46]See Somin, *Democracy and Political Ignorance*, Chap. 1, and discussion in Part I of this article.

[47]See, e.g., Ilya Somin, "Over 80 Percent of Americans Support Mandatory Labeling of Foods Containing DNA," *Washington Post*, Jan. 17, 2015, available at https://www.washingtonpost.com/news/volokh-conspiracy/wp/2015/01/17/over-80-percent-of-americans-support-mandatory-labels-on-foods-containing-dna/.

survey are almost purely technical questions. For example, the issue of whether GMO foods or foods treated with pesticides are safe, or the issue of whether human activity is the main cause of climate change.[48] On these sorts of technical matters, scientists are indeed likely to know much more than most ordinary people, and there is a good case for deferring to them. But some seemingly scientific policy issues actually include major nontechnical components on which scientists are not likely to have specialized knowledge.

B. The Limits of Scientific Expertise

Some of the questions raised in the Pew study are actually mixed questions of scientific facts and moral values. For example, the issue of whether animals should be used in scientific research partly depends on the scientific benefits of using them—a question on which scientists have special expertise. But it also depends on the moral status of the animals in question, and whether it is ethically permissible to inflict certain types of harm on them. On that latter issue, scientists have no special knowledge. If there is a group of experts that does, it is likely to be moral philosophers and political theorists; and these groups are—on average—more sympathetic to animal rights arguments than the general public is.[49]

Other issues on the survey raise questions of political economy rather than pure science. For example, many more scientists (82%) than ordinary people (59%) believe that growing population will be a "major" problem in the future.[50] Whether it will be or not depends largely on whether the possible costs of population growth (e.g.,—environmental externalities) will outweigh the benefits, such as increased innovation and a greater division of labor.[51] On these latter questions, economists are likely to be more expert than natural scientists are, and economists tend to be much more skeptical of Malthusian arguments than either natural scientists or the general population. They like to point out that Malthusian predictions have proven wrong for some two hundred years,[52] which does not prove that they will always be wrong, but does suggest reason for imposing a high burden of proof on them.

Even on issues when scientists really are expert, there is occasionally a case for discounting their views based on ideological bias, or narrow self-interest. For example, if we find that scientists are in favor of increased government subsidies for

---

[48]Pew Research CenterSurvey.

[49]Cf. Brian Leiter, "Philosophers, Eating, Ethics—A Discussion of the Poll Results," Leiter Reports, Oct. 10, 2012, available at https://leiterreports.typepad.com/blog/2012/10/philosophers-eating-ethics-a-discussion-of-the-poll-results.html. This survey is admittedly of low scientific validity. But it comports with conventional wisdom on the distribution of philosophers' opinions on this subject.

[50]Pew Research CenterSurvey, Jan. 29, 2015.

[51]See, e.g., Julian L. Simon, The *Ultimate Resource 2* (Princeton: Princeton University Press, 1996).

[52]See, e.g., Michael Shearmer, "Why Malthus is Still Wrong," *Scientific American*, May 1, 2016, available at https://www.scientificamerican.com/article/why-malthus-is-still-wrong/; "Malthus, the False Prophet," *The Economist*, May 15, 2008, available at https://www.economist.com/finance-and-economics/2008/05/15/malthus-the-false-prophet.

science, their position could be based purely on disinterested expertise. But it could also be special interest pleading, or some combination of both.

But it would be a mistake to dismiss all or most expert opinion on such grounds. Many of the issues on which experts and the public diverge have little direct connection to the self-interest of the former. Large lay-expert disagreements persist even in studies that control for self-interest and ideology, as Bryan Caplan did in his work comparing the views of economists and lay people on economic issues,[53] and we have in our joint work comparing the views of laypeople and political scientists on political influence (coauthored with Eric Crampton and Wayne Grove).[54]

In the case of the Pew survey, it is striking that scientists endorse what are usually considered "right wing" positions on nuclear power, GMO foods, and pesticides, even though scientists are generally much more left-wing in their political views than the average voter is.[55] The scientists could still be wrong about these issues. But if so it is not because of ideological bias.

In sum, it makes good sense to defer to the views of experts on areas that are actually within their expertise. But not on questions that may seem related, but actually are distinct. Telling the difference is not always easy. Here, as elsewhere, being a responsible, well-informed voter turns out to be a lot harder than we might think.[56]

Voters could improve their assessment of experts by searching out experts with opposing views and ideological commitments, considering their track records, searching indications of possible bias or conflict of interest, and checking whether the people in question are well-regarded by other experts.[57]

Unfortunately, in a world of rational ignorance, it is unlikely that many voters will act on such recommendations. Carefully assessing experts' claims, considering opposing views requires time, effort, and self-control that most people are unwilling to devote to the task.[58] Voters could devote substantial time and effort to considering experts' credentials and carefully evaluating the extent and limits of their expertise. But, thanks to rational ignorance, most are unlikely to do that.

---

[53] See Bryan D. Caplan, *The Myth of the Rational Voter*, (Princeton: Princeton University Press, 2007).

[54] Bryan Caplan, Eric Crampton, Wayne Grove, and Ilya Somin. "Systematically Biased Beliefs About Political Influence." *PS: Political Science and Politics* 46 (2013): 760–67.

[55] See, e.g., Pew Research Center Survey, July 9, 2009, available at https://www.pewresearch.org/wp-content/uploads/sites/4/legacy-pdf/528.pdf.

[56] For an overview of knowledge prerequisites for being a responsible voter under various normative theories of democratic participation, see Somin, *Democracy and Political Ignorance*, Chap. 2.

[57] For an overview of these and other similar approaches to assessing experts, see Alexander Guerrero, "Living With Ignorance in a World of Experts," in Rik Peels, ed., *Perspectives on Ignorance from Moral and Social Philosophy*, (London: Routledge, 2016); cf. Alvin Goldman, "Experts: Which Ones Should You Trust?" *Philosophy and Phenomenological Research* 63 (2001): 85–110; Harry Collins and Robert Evans, *Rethinking Expertise*, (Chicago: University of Chicago Press, 2007).

[58] For a review of some of these difficulties, see Guerrero, "Living with Ignorance in a World of Experts".

Similarly, voters could try objectively evaluate the views of contending experts, while holding their own biases in check. But that is far less emotionally satisfying than embracing information and seemingly expert opinion that reinforces preexisting views, while downplaying or just ignoring anything that cuts the other way. Here too, incentives to do better are weak.

The potential collective benefit of a smarter, more discriminating electorate is enormous. But there is very little incentive for individual voters to improve their performance. We should not rule out the possibility that there will be a major improvement in voters' use of information gathered from experts. But it seems unlikely that it will be achieved any time soon.[59]

5 How Foot Voting Can Help Us Make Better Decisions About Who to Trust

Is there a way to improve our incentives to seek out reliable information about political issues, and make good decisions about which information sources are trustworthy, and which ones not? There is no easy solution. But we can make progress by exploiting the informational advantages of voting with your feet over ballot box voting.

People can vote with their feet by choosing which jurisdiction to live in within a federal system, by making decisions in the private sector and through international migration.[60]

All three types of foot voting give decision-makers much stronger incentives to acquire relevant knowledge and evaluate it objectively than ballot box voting does.[61] If you are like most people, you probably spent more time and effort acquiring information the last time you decided which car, smartphone, or television set to buy than the last time you decided who to support in any election. Is that because the government of your country is less important than your TV, or deals with less complicated issues? Unlikely. It is because when people choose a TV, they intuitively realize that the decision is likely to make a difference. By contrast, when we vote for a presidential candidate, the chance that any individual decision will affect the outcome is infinitesimally small.

The same point applies to decisions on which jurisdiction to live in, decisions emigration, and decisions on private-sector transactions in the market and civil society.

---

[59]For a detailed discussion of various proposals to increase political knowledge, and explanations of why they are unlikely to succeed in the near future, see Somin, *Democracy and Political Ignorance*, Chap. 7.

[60]For extensive discussion of these three different types of foot voting, see Somin, *Free to Move*, Chaps. 1–4.

[61]I discuss this issue in much greater detail in Somin, *Democracy and Political Ignorance*, Chap. 5.

The key difference between foot voting and ballot box voting is that foot voters do not have the same incentive to be rationally ignorant as ballot box voters do. To the contrary, they have strong incentives to seek out useful information.

They also have much better incentives to objectively evaluate what they do learn. Unlike ballot box voters who can afford to be rationally irrational about political information,[62] foot voters know they will pay a real price if they do a poor job of evaluating the information they get.

That does not mean that foot voters are always well-informed or perfectly unbiased in their evaluation of information. Far from it. But, on average, they do a much better job than ballot box voters do.[63] This point is distinguishable from the long-standing idea that decentralization of political power can improve decision-making by concentrating more authority in the hands of governments that are "closer to the people" and deal with issues that voters are more likely to be able to observe directly.[64]

This general insight on the informational advantages of foot voting applies to the specific case of deciding which information sources to trust. Unlike ballot box voters, foot voters have strong incentives to acquire enough information to make an informed choice on these matters, while simultaneously attempting to evaluate claims to expertise in a relatively unbiased way.

These advantages of foot voting are well-illustrated by the debate triggered by recent works arguing that the rise of "fake news" and other quackery on the internet has undermined respect for genuine expertise, and made us more vulnerable to deception, such as Tom Nichols' *The Death of Expertise: The Campaign Against Established Knowledge and Why it Matters*[65] and Cass Sunstein's *#Republic: Divided Democracy in an Age of Social Media.*[66]

These problems are real. But Nichols oversimplifies by implicitly assuming that the problem of ignorance and contempt for expertise is largely uniform across issues. In reality, most of the really egregious examples he cites relate to political issues, not decisions made in the private sector and civil society. Many of the same voters who are ignorant of basic facts about government policy and angrily deny that expertise on such questions matters, also routinely go to doctors to cure their illnesses, hire mechanics to fix their cars, and follow expert advice on many other aspects of their

---

[62] See discussion in Part I.

[63] For a variety of evidence to this effect, see Somin, *Democracy and Political Ignorance*, Chap. 5.; for studies showing that people analyze political information in a more biased way than other information, see e.g., Dan M. Kahan, Ellen Peters, Erica Cantrell Dawson, and Paul Slovic, "Motivated Numeracy and Enlightened Self-Government," Behavioral Public Policy 1 (2017); 54–86; Dan Kahan, "Misconceptions, Misinformation, and the Logic of Identity-Protective Cognition," SSRN *Electronic Journal*. https://doi.org/10.2139/ssrn.29730 67.

[64] For a recent argument to the effect that this can reduce the impact of political disinformation, see O'Connor and Weatherall, *The Misinformation Age*, 176–77.

[65] Tom R. Nichols, *The Death of Expertise: The Campaign Against Established Knowledge and Why it Matters*, (New York: Oxford University Press, 2017).

[66] Cass R. Sunstein, *#Republic: Divided Democracy in an Age of Social Media*, (Princeton: Princeton University Press, 2017).

lives. The same people who fall prey to political hucksters and demagogues who skillfully use the internet to their advantage, have little difficulty figuring out that you get cheaper and better products at, say, Amazon, than on some shady fly-by-night website.

Outside the realm of politics, it is difficult to deny that the benefits of online information flows vastly outweigh their costs. The internet has enabled people to access an incredible range of better and cheaper goods and services compared to what existed before. Internet-based businesses such as Uber, Lyft, and AirBnb even allow us to entrust our travel, comfort, and safety to complete strangers secure in the knowledge that we are likely to get better and cheaper service than through their more conventional competitors. There is no shortage of nonpolitical con artists and hucksters online. But most internet consumers have learned to avoid them, or at least minimize the risk.

By contrast, we do a much worse job of minimizing the risk posed by political deception and disinformation. If you get an e-mail from a wealthy heiress who offers to pay you a million dollars, or see a website that promises to increase your sexual potency at a bargain-basement price, you are likely to be highly skeptical. By contrast most voters people do not apply anything like the same degree of common-sense scrutiny to political snake oil—especially when it conforms to their preexisting views. Politicians and activists who peddle dubious conspiracy theories, promise to give you something for nothing, and otherwise spread disinformation, often gain a wide following.

Throughout *#Republic*, Sunstein argues that we should seek ways to get people to act like "citizens" rather than mere "consumers."[67] The latter merely want to satisfy their existing preferences, including the preference for information that conforms to their preexisting viewpoints. By contrast, citizens care about the public interest, and are willing to reexamine their preferences, and give opposing views more of a hearing.

There is some merit to this perspective. But Sunstein overlooks another important way in which consumers differ from citizens. Consumers are foot voters who make decisions they know are likely to make a difference. That gives them stronger incentives to seek out information and consider it at least somewhat objectively. By contrast, when we are in "citizen" mode, we may indeed care about the public interest. But we also know that our choices are unlikely to make much difference, and therefore make very little effort to either assimilate information or ride herd on our prejudices.

Perhaps, therefore, the antidote to the perils so effectively outlined by Nichols and Sunstein is not less consumerism, but more. We should work to ensure that more of our decisions are like the ones we make about smartphones and cars, and fewer have the perverse structure of incentives that exists at the ballot box.

The informational advantages of foot voting over ballot box voting can be more effectively exploited if we take steps to limit and decentralize government. The more decentralized government is, the more issues can be decided through foot voting.

[67] See Sunstein, #Republic, especially Chaps. 6, 9, 11.

It is much easier to vote with your feet against a local government than a regional government (except, perhaps, a very small regional), and much easier to do it against a region than against the national government.

It is also usually much easier to foot vote in the private sector than the public. A given region is likely to have far more private planned communities and other private sector organizations than local governments.[68] Choosing between the former usually requires far less in the way of moving costs than choosing between the latter.

Foot voting has significant downsides as well as upsides. In previous work,[69] I cover several standard objections to foot voting, such as the problem of moving costs, the danger of "races to the bottom," and the likelihood that political decentralization might harm unpopular racial and ethnic minorities. Each of these concerns is sometimes a genuine problem. But I contend that each one is a less severe challenge than commonly believed. For example, moving costs can be reduced by decentralizing to lower levels of government or to the private sector, and are in any case declining thanks to modern technology.[70]

Mitigating the dangers of political ignorance and bias is, of course, far from the only factor that must be considered in deciding the appropriate size, scope, and centralization of government. But it deserves greater consideration in debates over the role of government in society than has so far been the case.

6 Conclusion

Neither trust in information sources nor distrust is intrinsically good or bad. The goal should be to find the right balance between the two. We should not seek any particular level of trust in information sources, generally. Rather, the goal should be to expand trust in good information sources, while promoting deserved skepticism about more dubious ones.

Ideally, voters could judge the quality of politically relevant information entirely on their own. But, given real-world constraints, some degree of deference to expert judgment is unavoidable. There is no simple way to get the balance right. But understanding the dynamics of rational ignorance and irrationality can help us begin to see how to make things better. Appreciating the relevance of the informational advantages of foot-voting is another useful step.

The longterm objective should be to altr the perverse incentives that lead voters to make poor decisions. Reducing our reliance on ballot-box voting and making greater use of foot voting can help us move in the right direction.

---

[68]On the advantages of private planned communities as mechanisms for foot voting, see Somin, *Free to Move*, Chap. 4.

[69]See ibid., Chaps. 2–6, and Somin, *Democracy and Political Ignorance*, 165–76.

[70]See Somin, *Free to Move*, chap. 2; Somin, *Democray and Political Ignorance*, 165–67.

Trust and Fiduciary Power

Marco Menon

Abstract This paper analyzes an aspect of trust relatively less explored by political philosophers, namely, the legal conception of trust. As I explain in Sect. 1, the legal conception of trust refers to trust as a moral and/or legal obligation of power. It is distinct from trust understood as confidence and from trustworthiness; nonetheless, I offer that trust as confidence is constitutive of trust as obligation. In Sect. 2, I briefly recount the origin of the trust and the development of the legal metaphor of trust in 17th century England, most notably in the work of John Locke. In Sect. 3, I turn to examining the contemporary fiduciary political theory proposed by Evan Fox-Decent, which is a literalist transposition of private fiduciary principle to the public sphere. One notable aspect of Fox-Decent's theory is its attempt to justify the State's legitimacy on the basis of its fiduciary relationship with its subjects. Fox-Decent thus offers an alternative to all consent-based theories, primarily contesting A. John Simmons's voluntarism. As I explain in Sect. 4, fiduciary political theory is based on a "presumption of trust" and appeals to Annette Baier's conception of an "unconscious, automatic" trust in order to explain the fiduciary relation between the individual and the irresistible administrative power of the State. In Sect. 5, I present the main objections to fiduciary political theory and explore the compatibility of the fiduciary principle with Simmons's voluntarism.

Keywords Trust · Fiduciary power · Legitimacy · Justification · Political obligation · Fiduciary principle

M. Menon (✉)
University of Pisa, Department of Civilizations and Forms of Knowledge, Pisa, Italy
e-mail: marco.menon@cfs.unipi.it

A. Fabris (ed.), *Trust*, Studies in Applied Philosophy, Epistemology and Rational Ethics 54, https://doi.org/10.1007/978-3-030-44018-3_12

1 Introduction

The aim of this paper is to offer a politico-philosophical perspective on a conception of trust that, in the ever-growing literature on the subject, has remained "strangely peripheral to the social science literature" (Braithwaite 1998, 344)[1]: the legal conception of trust, its role in the history of political philosophy, and its contemporary developments in political theory. At face value, this statement may appear highly controversial and therefore needs some clarification. Is it not the case that the legal conception of trust is related to "common-sense" trust only in name? One may object from the outset that the legal conception of trust is not a lesser explored aspect of trust broadly understood. In fact, it appears to be something else. But, as will hopefully become clearer in the following, the two notions are strictly connected. In order to clarify the nature of the subject matter, we can follow Braithwaite again and note that "[t]rustworthiness is the social science conception that comes closest to this legal conception," concluding that "[t]he dominant social science conception is of trust as confidence, as opposed to trust as obligation." In fact, the legal conception of trust refers to "trust as a moral obligation of power" (Braithwaite 1998, 344), namely, the fiduciaries' (as wielders of some kind of other-regarding power) obligation to live up to the trust placed in them by those who are affected by their power. When such a power is the administrative power of the State, some political theorists imply, then such an obligation binds the governors and the public officers. The State (or government) then can be understood as the fiduciary of the people. It is in this sense that I will refer here to legal trust in its political declension.

The following is an outline of a politico-philosophical treatment of the concept introduced above. I think such a treatment is needed for at least three reasons. Firstly, as recounted (briefly) in Sect. 2, the legal metaphor of trust (inspired by the trust of the law of equity, and heavily conditioned by the common notion of trust as confidence) has played an important role in the history of modern political thought and exerted decisive influence on Western political theory by virtue of John Locke's

[1] Braithwaite's statement finds further confirmation in the work of J. S. Maloy. As the latter summarizes (2009), the recent literature on trust produced by political philosophers and scientists, when considering the specifically political dimensions or aspects of trust, has analyzed and examined the subject by mainly assuming a "psychic approach." Authors mostly see trust as a "psychic quantity," which becomes "political" when it regards institutions and politicians, or the relationship between politicians and ordinary citizens, or between the citizens themselves. The "psychic approach" also implies an "economic" understanding of trust, in the sense that it is seen "as a kind of resource amenable to increase, conservation, depletion, etc." (Maloy 2009, 493). This approach then splits, according to Maloy, into two major diverging orientations: on the one hand, the "civic" or "social capital" line, epitomized by Robert D. Putnam; and on the other, the "legalistic" or "game-theoretical" line, epitomized by Russell Hardin. While the civic orientation tends to focus on the relationships between citizens and the social practices of mutual and/or generalized confidence that support healthy democratic life, the legalistic orientation, conversely, tends to focus on the relationship between citizens and institutions, and considers the ways in which institutions must be conceived in order to be supportive of general trust. In other words, according to the legalistic orientation, institutions represent the conditions "under which it is rationally justifiable to trust" (Maloy 2009, 493).

notion of fiduciary power. Secondly, in the last years we have seen the birth and constant growth of fiduciary political theory, a position elaborated mainly by public and private law scholars and based on fiduciary principles of private law. Most notably the efforts by Evan Fox-Decent, Evan Criddle, Tamar Frankel, Andrew Gold, Ethan J. Leib, Paul Miller, and Stephen R. Galoob have shown the great flexibility of the fiduciary principle and its potential application to fundamental problems of political philosophy (above all, the problem of political obligation), as will become evident in Sects. 3 and 5. Thirdly, the metaphorical rendering of legal trust, as well as its literal application to the public sphere, are based on a more common understanding of trust as confidence, even though in different ways. In particular, as shown in Sect. 4, the fiduciary political theory proposed by Fox-Decent is based on a peculiar conception of trust introduced by Annette Baier, namely, "unconscious or automatic trust." In this way, I will show how a philosophical elaboration of the concept of trust has been adopted by a public law theorist in order to construe a new fiduciary political theory.

2 Fiduciary Power: Between Legal Metaphor and Political Obligation

The debate concerning the genealogy of trust is very intricate and transcends the scope of the present discussion. This notwithstanding, it seems necessary to at least provide a succinct account of the historical development of legal trust. The trust's medieval forerunner, the use or *feoffment to use*, was a practice devised mainly to dodge a landholder's feudal obligations to his lord (i.e., the lord's rights of wardship; for a detailed reconstruction, see Seipp 2011). The use, which shows correspondences with the Roman *fideicommissum*, the German *treuhand* or *salman*, and the Islamic *waqf* (Ames 1908; Gvelesiani 2016), consisted in the transfer of the legal property of a land, provided the beneficial enjoyment was not granted to the new legal owner, but to a third, the *cestui que use*. Common law did not recognize any right of the *cestui que use*, and the persistence of the agreement could rely only on the trustworthiness of the legal owner, or *feoffee*. The trust strictly understood emerged in the context of the law of equity, after Henry VIII, with the introduction of the *Statute of Uses* (1536), identified the *cestui que use* as the legal owner, thereby "nullifying the role of the fiduciary" (Luisi 2016, 53n5). Equity was that body of law enforced by the Court of Chancery and distinguished as such from common law. People who, for various reasons, could not find justice at courts of common law (Common Pleas, Exchequer, King's Bench), appealed to the Court of Chancery, which was meant to redress the typical harshness of the rigid common law. To the Chancellor's jurisdiction belonged all trusting and fiduciary relationships, included the medieval use and its successor, the trust.

Like the use, a trust is not a contract, but an arrangement involving three parties based on confidence and a duty of trustworthiness binding in conscience. The three parties are the settlor or entrustor, the trustee or fiduciary, and the *cestui que trust* or beneficiary. The settlor entrusts a property to the trustee; by virtue of this grant, the trustee becomes the legal owner of that property, but he is bound to exercise his property rights in the interest of another person (the beneficiary), according to the terms of the agreement. Using the words of Frederic W. Maitland, one could formulate a definition of the trust as follows: "When a person has rights which he is bound to exercise upon behalf of another or for the accomplishment of some particular purposes, he is said to have those rights in trust for that other or for that purpose" (Maitland 1910, 43).

Between the 16th and the 17th century, when the trust was imposing itself as a familiar institution in England, the usage of the term "trust" as a political concept progressively became popular in public debate, and backed up the affirmation of popular sovereignty (for a comprehensive overview, see Gough 1950).[2] On the superficial level, talk of trust in politics implied an extension of "common-sense" trust into the public sphere. The people trusted those in office to do their duty with loyalty and care. But the term was also used in a more technical sense, referring metaphorically to the trust of the law of equity. Political power was held in trust by the King and the parliament *as if* they were legal trustees or fiduciaries of the people. This specific usage indicates that political power can neither be absolute, nor arbitrary, and must be accountable to the people (Gough 1950, 138 ff; Dunn 1984, 296; Laslett 1988, 113 ff). In hindsight, the author that gave this usage a paradigmatic doctrinal exposition was John Locke, whose political theory was to become highly influential in the following centuries, on both sides of the ocean. In the *Second Treatise of Government*, Locke elaborates a sophisticated account of the origin of the political power detained by the government. On one hand, he conceives of the supreme power, granted by the community to the legislative, "only" as a "fiduciary power to act for certain ends" (Locke 1988, 367). On the other hand, the executive's power is a "fiduciary trust placed in him, for the safety of the people" (Locke 1988, 371). As noted by John Dunn, Locke employs the "metaphor of a legal trust" (Dunn 1984, 296), while according to Peter Laslett, he makes "suggestive use of legal language" (Laslett 1988, 115). It is recognized that Locke makes free use of that institution, plausibly drawing also on the Roman notion of *fides publica* (de Wilde 2011), for the dual political relationship between the community and the government does not match the three-parties agreement of the trust. In his scheme, the community is the entrustor and the government is the trustee—but the beneficiary is the community itself. By doing so, Locke conflates the entrustor with the beneficiary, reducing a three-parties legal arrangement to a two-parties political relation. Even though his usage of the notion of legal trust is only metaphorical, it remains highly significant because his intention was to distinguish the "social contract" from the "governmental trust." The latter, in fact, implies a different attribution of rights and duties to the parties in the arrangement. In fact, in the first place, a fiduciary cannot act in his own interest. He must exercise his discretionary power only in the best interest of the beneficiary, and according to the end established. In the second place, while the parties of a contract each have rights of their own, in a

[2]Previously it was assumed that "offices granted *by the king* were held 'on trust and confidence'" (as stated in a Year Book of 1471, quoted in Seipp 2011, 1025, my emphasis).

trust relationship "the trustee does not enter into a contract" with the beneficiary, for "he simply accepts an obligation, and accepts it unilaterally" (Barker 1962, xxiii; see also Simmons 1993; Nacol 2011). In the case of Locke's adaptation, the beneficiary can be labeled a "dormant" entrustor. In this way, the latter holds the residual right or power to "discard" the trustee "when he fails in his trust" (Locke 1988, 427).[3]

Locke's teaching was a prudent or moderate synthesis of 17th century radical Whig political thought (Maloy 2009). His conception of political trust influenced in a decisive way many important 18th century political thinkers, who construed the relationship between rulers and ruled in the model set out in the *Second Treatise of Government*. In this respect, a few examples will suffice. Thomas Gordon and John Trenchard, authors of *Cato's Letters*, insisted that "government is a trust" (Zuckert 1994, 314). Edmund Burke understood government as a trusteeship and, notably, conceived of the Indian people as the beneficiary of a trust relation in which the fiduciary was the British East India Company, and the trustor was the English Parliament (together with the Great Mogul).[4] John Adams held that "[r]ulers are no more than attorneys, agents, and trustees, for the people; and if the cause, the interest and trust, is insidiously betrayed…the people have a right to revoke the authority that they themselves have deputed" (Adams 1851, 456–57).[5] The notion of fiduciary power or trust was shared also by such thinkers as Jeremy Bentham and John Stuart Mill,[6] but later, especially in the 20th century, it progressively receded to the background of political thought and political philosophy, becoming almost exclusively relevant in the sphere of private property.

3 Beyond Metaphor: Fiduciary Political Theory

After decades of desuetude, the notion of the State as fiduciary reappeared in the academic debate in 1995 thanks to Paul Finn's *The Forgotten Trust: The People and the State*, an article now recognized as one of the milestones of the so-called "fiduciary political theory".[7] To use the words of some critics, fiduciary political theory "is an intellectual project that seeks to recover the fiduciary foundations of public

---

[3]So, the community is the wielder of actual sovereign power, as noted in Knoll (2010, 236). It is also for a similar reason that Jean Hampton proposes that what Locke had in mind was not a trustee/beneficiary relationship, but a principal/agent one (see Hampton 1986, 248n18). The latter is a fiduciary relationship based on a contract in which the principal controls the agent (see also Pettit 2012, 32).

[4]On Burke, see Conniff (1977, 1993).

[5]On this point see Foa Dienstag (1996).

[6]"Sovereign power…upon the principle of utility, can never be other than fiduciary" (Bentham 2005, 263). "In whatever way we define or understand the idea of a right, no person can have a right…to power over others: every such power, which he is allowed to possess, is morally, in the fullest force of the term, a trust" (Mill 1998, 353–54).

[7]The term was introduced by Leib et al. (2013, 91); on Finn's contribution see Leib et al. (2014, 396); for a comprehensive theoretical introduction see Frankel (2011).

authority. It takes seriously the idea that political office is a public trust, which must be administered seriously with sensitivity to the implied fiduciary role officers take on." Its fundamental idea is that "private law fiduciary principles underlie the ethics of public office" (Leib et al. 2014, 388). According to Galoob and Leib, it is possible to distinguish between at least three different approaches to fiduciary political theory, based on the way in which the theorists conceive of the applicability of private law principles to public law: (1) The literalist approach "sees political relationships as instantiations of fiduciary relationships" (Galoob and Leib 2018b, 413). The followers of this approach recognize the same normative fiduciary structure in both private and public relationships, "literally" apply private law principles to public law, and, at least in one notable case, propose to use them to solve the fundamental problem of political obligation. (2) The analogical approach recognizes, in certain public-domain relationships, close resemblances to private fiduciary relationships. On the basis of these resemblances, the public domain should be bound by the same fiduciary constraints that apply to private relationships. (3) The translational approach "applies fiduciary norms to public domains through interpretation, rather than directly or by analogy" (Galoob and Leib 2018b, 414), to the effect that fiduciary principles are first identified, then adjusted to the *sui generis* character of the public sphere, and finally used to illuminate its crucial aspects.

In the present discussion, I will focus on the contribution of Evan Fox-Decent, who adopts a "literal" approach to fiduciary theory and appeals to the fiduciary principle in order to solve a fundamental issue of political philosophy. To date, his most notable solo effort is represented by the book *Sovereignty's Promise: The State as Fiduciary* (Fox-Decent 2011). The fiduciary theory defended therein is wide in scope and ingenious in its philosophical argumentation. It aims at a thorough interpretation of the State-subjects relationship in fiduciary terms, for it maintains that the State is the trustee of its subjects, and therefore bound by peculiar fiduciary legal obligations. This theory intends to put on display a global explanatory capacity, for "the fiduciary view of authority is fractal in the sense that it retains the same structure on any scale," a feature which enables it to "explain (non-consensual) authority on any scale" (Fox-Decent 2011, 48). Its range extends from paternal power to the State's political authority. Both relationships involve some kind of non-consensual authority which can be understood only in the light of the fiduciary principle. Even though this fractal character of the theory does not imply the identification of paternal power with political power, it hints at a structural correspondence between the two. In this respect, Fox-Decent's position is at odds with Locke's radical distinction between paternal power and civil government. But more fundamentally, it distinguishes itself from Locke's conception of fiduciary power for the fact that Fox-Decent's notion of public trust is not metaphorical. In his account, the state is literally and, therefore, legally the trustee of its subjects (citizens and residents alike). His scheme recovers the three-parties private law arrangement of the trust and transposes it, with some subtle adjustments, into public law terms: the law is the entrustor or settlor, the State is the trustee or fiduciary, and the people are the beneficiary. The power held by the State is "of an administrative nature," that is, "always other-regarding, purposive, and institutional in nature" (Fox-Decent 2011, 30) and exercised on behalf of the beneficiaries. The latter (i.e. the subjects) are vulnerable and constitutively unable to control the power of their fiduciary.[8] Instead of the moral and political obligation envisioned by Locke and his followers, according to Fox-Decent's theory the State owes its people the fulfillment of specific *legal* fiduciary duties, first of all what corresponds to the private trust duty of loyalty, namely, a duty of fairness and reasonableness, since the public is composed of multiple beneficiaries (Fox-Decent 2005, 265; Fox-Decent 2011, 37).

A particular feature of this fiduciary political theory is its legitimating claim. This aspect is not immediately apparent. As the author declares, "the most that can be expected of the common law fiduciary principle is that it can impose a legal standard where discretion would otherwise prevail" (Fox-Decent 2011, 76). In fact, this theory looks at the States "regardless of their origins," be they consensual, violent, revolutionary, elitist, and so on. What matters most is the State power's "present nature and consequences" (Fox-Decent 2011, 86), and the measure in which it lives up to its task as fiduciary of the people by enforcing the rule of law. Take any actual State, Fox-Decent seems to say, and I will assess its legitimacy according to a fiduciary standard of legality, which implies "a single but complex legal obligation, one which arises solely from the fiduciary character of its relationship to the people" (Fox-Decent 2011, 3). In this sense, his theory starts from the brute "fact of sovereignty," (Fox-Decent 2011, 29, 111) which as such seems to be the result of a unilateral initiative (comparable, *mutatis mutandis*, to the unilateral act of procreation). What indeed cannot be unilateral at all are the legal and moral terms of the relationship originated by the institution of a de facto sovereignty. These terms are set by the fiduciary principle, that is, by the law. "Since the law does not permit one part to set unilaterally the terms of its interaction with another, the fiduciary principle intercedes *when one party unilaterally assumes discretionary power over another* to ensure that the power holder has due regard for the best interests of the beneficiary" (Fox-Decent 2011, 68, my emphasis).

I am convinced that Fox-Decent's theory deserves particular attention from political philosophers and political theorists. I will justify this statement by concentrating on a single but significant aspect, namely the problem of political obligation, or the moral duty to obey the law. Fox-Decent challenges all consent-based theories of political obligation, above all A. John Simmons's Lockean voluntarism,[9] and proposes an alternative line of argument. His comprehensive fiduciary theory offers a basis for the justification of legal authority (which is almost the equivalent of what Simmons calls "legitimacy," cf. Simmons 2001 and Fox-Decent 2005, 290).[10] In Fox-Decent's

---

[8]A comparison should be made with Philip Pettit's contestatory republicanism (Pettit 1997). While Fox-Decent's idea of the subject's legal inability to exercise or control the administrative power of the State seems to be at odds with Pettit's approach, the same cannot be said about Fox-Decent's common law constitutionalism as a republican guarantee against domination (cfr. Fox-Decent 2011, 26).

[9]Voluntarism is a political theory that grounds political obligation on express consent alone; see Simmons (1981).

[10]Simmons distinguishes between justification and legitimacy of the State in his classic article *Justification and Legitimacy*. According to Simmons, "we can justify the state by showing that

scheme, "the state's *legal* authority" is "the authority to make, interpret, administer and enforce law," and it must be distinguished from the state's political authority, namely "the authority to determine the substantive character of ordinary law within the constitutional framework supplied by the state's legal authority" (Fox-Decent 2005, 262). It is only this latter that is based on consent. Ultimately, the state's legal authority amounts to "the power to create morally binding legal rules" (Fox-Decent 2011, 115). Fox-Decent's argumentative strategy amounts to a reply to Simmons' theory in Simmons' own terms. He argues that consent cannot be the basis of political obligation for the following reasons: Consent is either hypothetical or actual. If consent is hypothetical, "possible consent to some possible state does not entail actual consent to any particular state," to the effect that "it is not clear why I owe any obligation to *this* state rather than some other" (Fox-Decent 2005, 288–89). If, alternatively, consent is actual, we would enter a slippery slope that leads us to philosophical anarchism, a position Fox-Decent discards. Ultimately, at the basis of philosophical anarchism would stand Simmons' allegedly incorrect assumption that consent is the sole conceivable basis of legitimacy. In truth, "consensual relationships are not the only kinds of relationships in which one may come to have authority over another" (Fox-Decent 2011, 121).

Notwithstanding his radical criticism of Simmons, Fox-Decent adopts his interlocutor's argumentative structure, for he recognizes that the basis of the state's legal authority/legitimacy must be "an appropriate moral relationship between state and subject" (Fox-Decent 2011, 120). This is also Simmons's fundamental assumption. The point is that, according to Fox-Decent, that particular moral relationship cannot be a contract-like relation based on consent, but a fiduciary relation based on trust, and premised on the Kantian conception of innate rights.

---

some realizable type of state is on balance morally permissible (or ideal) and that it is rationally preferable to all feasible nonstate alternatives" (Simmons 2001, 125–26), while "[p]olitical power is morally legitimate, and those subject to it are morally obligated to obey, only where the subjects have freely consented to the exercise of such a power and only where that power continues to be exercised within the terms of the consent given" (Simmons 2001, 129). Simmons specifies that "State legitimacy…includes an exclusive power over subjects to impose duties and enforce them coercively, which correlates with obligations on others to refrain from these tasks. It also includes a right, held against subjects, to be obeyed…This latter right is the logical correlate of subjects' political obligations" (Simmons 2001, 130n20). It must be noted that in a previous article, Fox-Decent specified that his "claim regarding the legitimacy implied by state legal authority is much narrower in scope and ambition than the all-things-considered legitimacy that Simmons addresses…My project concerns solely the legitimatizing effects of legality…a state's compliance with the fiduciary view of the rule of law gives the subject *a* reason to obey the law" (Fox-Decent 2005, 290, emphasis in the original). In his later *magnum opus*, Fox-Decent seems to offer a more radical view.

4 Fiduciary Relations and the Presumption of Trust

The moral basis of Fox-Decent's theory is, hence, twofold. He establishes his fiduciary theory on "a moral idea of trust premised on a Kantian conception of right." This combination explains "both the authority and obligation under which a fiduciary exercises power" (Fox-Decent 2011, 46); that is, both the legal authority of the State and its fundamental fiduciary duties of fairness and reasonableness. This part of Fox-Decent's fiduciary theory is perhaps the most fascinating for political philosophers. The first move of its argumentative strategy is the adoption of a Kantian conception of an innate right to equal freedom by virtue of our shared humanity. Every individual is "a being who possesses an innate moral capacity to place others under obligation" (Fox-Decent 2011, 105). The requisite of equal freedom implies the illegitimacy of unilateralism, to the effect that no one has the right to impose unilaterally the terms of a relationship. The fiduciary principle establishes itself on this moral ground, for it constrains every unilateral assumption of power to respect a fiduciary duty of loyalty (fairness and reasonableness) when the latter influences the practical interests of vulnerable subjects. This is where the second move comes in, which brings into play the notion of "trust" in its common meaning of confidence. In Fox-Decent's scheme, the beneficiary does not need to consent to the fiduciary's exercise of power on her behalf. The latter's legitimacy is grounded not on a contractual-like relationship based on consent, but on a trust-like relationship based on trust, to the effect that "when a fiduciary exercises irresistible administrative power, she does so *on the basis of the beneficiary's trust*" (Fox-Decent 2011, 105, my emphasis).

At this point, Fox-Decent's proposal sounds counterintuitive, for our commonsense notion of trust as confidence in someone's good disposal toward us seems to imply a certain degree of consent. In the typical dynamic relationship of trust, if A entrusts herself to B, B acquires some measure of discretionary power over A; since trust is given freely, it is commonly assumed that A has freely consented to make herself vulnerable to B. How does this fit Fox-Decent's statement, according to which the fiduciary relationship is based on the beneficiary's trust and *not* on her consent? In order to prove his point and to show the plausibility of an argument for political obligation which dismisses consent as unnecessary, Fox-Decent turns to a peculiar trust relation, drawing on Annette Baier's classic paper "Trust and Antitrust" and her notion of an "automatic and unconscious" trust. Baier states that,

> [w]hereas it strains the concept of agreement to speak of unconscious agreements and unchosen agreements, and overstrains the concept of contract to speak of unconscious or unchosen contracts, there is no strain whatever in the concept of automatic and unconscious trust, and of unchosen but mutual trust. Trust between infant and parent, at its best, exhibits such primitive and basic trust (Baier 1986, 244–45).

Fox-Decent finds this distinction of Baier's extremely useful to his project, for the infant's trust towards the parent resembles the fiduciary relationship between the subject and the State, even if that does not reduce the subject to an infant and the State to a parent. What matters most is the fact that the "automatic and unconscious" trust exhibited by the infant-parent relation "may capture the sense of trust that people have in the state that exercises such power over them, even if they have never done anything explicit to repose trust in the state" (Fox-Decent 2011, 108). Trust, in the sense just presented, substitutes consent and seems to overcome all of the paradoxes in which any consent-theory must incur. For the "automatic and unconscious" trust is actual trust and connotes decisively the significant moral relation between subjects and the State. So, if "we can act on the basis of another party's trust even if that party has done nothing to repose trust in us," "it follows that a fiduciary can act on the basis of beneficiary's trust even if the beneficiary has done nothing to repose trust in the fiduciary" (Fox-Decent 2011, 108–9).

As we have seen, in the scheme contrived by Fox-Decent, the beneficiary does not need to express her trust to let the fiduciary assume this trust as real and act accordingly upon such an assumption. The fiduciary, who exercises an irresistible administrative power over the vulnerable subject's interests, can act legitimately on her behalf on the basis of a "presumption of trust" (Fox-Decent 2011, 93). But, in truth, "it is not the beneficiary actually reposing trust in the fiduciary that does the work, but rather the fiduciary principle itself that requires the office-holder to act on the basis of the principal's trust" (Fox-Decent 2005, 263). In order to explain this point and offer a plausible model of the State's legal authority based on such a "presumption of trust," Fox-Decent appeals to the legal notion of "agency of necessity."[11] This notion allows him to conceive of the State as a "public agent of necessity" which "possesses and exercises public power that private parties are not entitled to exercise" (Fox-Decent 2011, 49). The agency of necessity is justified by the juridical "inability on the part of the principal to exercise the power that law has entrusted to the agent" (Fox-Decent 2011, 135). The administrative power of the State as "agent of necessity," not based on consent by the beneficiary but on a presumption of trust, overcomes the problem of unilateralism, provided it acts in the full respect of its fiduciary obligations and establishes the rule of law.

Hence it appears quite clear that what really matters in Fox-Decent's scheme is not so much any actual trusting on the part the beneficiary, be it automatic or unconscious, but rather the fulfillment, required by the law, of certain moral-and-legal obligations emerging in a fiduciary relationship. In other words, the law commands the de facto sovereign, in order to become legitimate, to act on the basis of the duties connected to a trust-like relationship with its subjects. It is only on the basis of such an assumption (and fulfillment) of duties, imposed by the fiduciary principle, that a de facto sovereignty becomes a *de jure* sovereignty. By acting on the basis of fiduciary obligations, the State acquires legal authority, that is, the moral right to be obeyed by its subjects. The moral justification of the duty to obey the State can be thus summarized as follows: On one hand, the Kantian conception of an innate right to equal freedom and non-domination places the State under the obligation to guarantee legality, equal liberty, and security to its subjects. On the other hand, the fiduciary

[11] "The clearest case of agency of necessity involves shipmasters who find themselves *in an emergency situation* that places the cargo they are carrying in imminent peril…Shipmasters normally have no contract with the owners of the cargo. Nonetheless, courts have held that the shipmaster may act *without prior authority* as an agent of the cargo's owner in order to protect the goods or their value" (Fox-Decent 2011, 132, my emphasis).

principle, by imposing fiduciary obligations on the State's unilateral assumption of irresistible administrative power, justify its legal authority and thereby grounds the subjects' moral duty to obey the law. Fox-Decent's entire construction intends to show that consent is not required for the legitimacy of the State. It aims at a de facto refutation of Simmons' voluntarism.

5 Issues and Further Applications of the Fiduciary Principle

This theoretical effort has aroused a lively debate about the possibility to apply private law principles to public law and political theory, and the ways in which such an application can be effectively achieved. In this context, I will concentrate on the main objections to this legitimizing project, and briefly explore the compatibility of the fiduciary principle with a "Lockean" consent-oriented theory of political obligation.

The legitimizing performance of the fiduciary principle faces two main objections. The first one, philosophical in character, concerns the effective achievements of the fiduciary principle. It seems, in fact, that the State's compliance with fiduciary duties, as required by the law, and its exercise of irresistible administrative power according to the standard of legality elaborated by Fox-Decent, gives us "*a* reason to obey the law," yet not a moral duty to obey it. In truth, as Fox-Decent recognized in an earlier exposition of his theory, "legality is a necessary but not sufficient condition of all-things-considered legitimacy" (Fox-Decent 2005, 290; emphasis in the original; "all-things-considered legitimacy" here refers to legitimacy as understood by Simmons). The claim raised in his later systematic work is clearly more ambitious. But the earlier admission seems to be truer to the effective legitimizing capacity of the fiduciary principle. According to a critic, Fox-Decent's later exposition seems to conflate "the fact that compliance with fiduciary constraints legitimates exercise of fiduciary authority with the distinct (and untenable) supposition that the existence of fiduciary constraints can legitimate a claim to possession of fiduciary authority" (Miller 2016, 254). In Simmons's terms, the fiduciary principle would greatly contribute to the State's justification, but not to its legitimacy. The citizens would have significantly good reasons to support, or at least not to interfere with, Fox-Decent's ideal State; still, it is doubtful that they would be morally obligated to obey it on the basis of the fiduciary principle alone.

The second objection concerns Fox-Decent's appeal to the notion of "agent of necessity," and is of a purely legal character. As noted by Andrew Gold, among others, the form of agency Fox-Decent appeals to in order to characterize the fiduciary relation between State and citizens does not really fit this last kind of relationship. The requirement of the agency of necessity is the impossibility of communication between the agent and the principal. Therefore, the appeal to this legal institution "might suggest that the state has legitimate authority with respect to children, individuals in a coma, or individuals who are mentally incompetent. Extending the agency of necessity doctrine to the public setting, the state's legitimate authority would be quite limited" (Gold 2013, 667). It is evident, though, that the legal incapacity of private subjects asserted by Fox-Decent would encourage consideration for the State as an agent of necessity. But this seems to be equally questionable, for "agents of necessity don't generally acquire the ability to order their principals around" (Gold 2018, 187). It must be added that Fox-Decent's theory has also been criticized on the basis of its "literal approach," namely, for applying the fiduciary principle to public law and political theory not on the basis of a structural analogy between the two domains, nor on the basis of a "translation," but in virtue of the essential identity of the fiduciary/beneficiary relationship in private law and in public law.

Notwithstanding the criticism briefly summarized here, there are aspects of fiduciary political theory that may offer reasons for further research to political philosophers and political theorists alike. I will focus on two points: firstly, its compatibility with a "Lockean" consent-based approach; secondly, its potential contribution to a moral code of governors as fiduciaries.

(1) Several authors recommend a partial opening of the fiduciary camp to consent as the basis of political obligation, for the reason that the fiduciary principle alone does not seem capable of legitimizing the State's legal authority. In light of this, it appears necessary to renounce to Fox-Decent's comprehensive view and pursue a different strategy. On the basis of the compatibilist approach advanced by Galoob and Leib (2018a), it seems plausible to combine the fiduciary principle with different theories of legitimacy, among them the very voluntarism Fox-Decent so strenuously opposes. Galoob and Leib also maintain that Simmons' theory would need a certain modification in order to become hospitable to the fiduciary principle. But I would say that his radical voluntarist scheme already provides a footing to it. I am referring to those non-consensual interactions in which a party is allowed to normatively affect the condition or interests of another party. In fact, it is not fully correct to assert that, in Simmons' view, "citizens' volitional commitments are both *necessary and sufficient* to explain a state's right to rule over them" (Galoob and Leib 2018a, 180, my emphasis). Certain normative interactions can be non-consensual and still be morally binding; namely, those commanded by natural law. As Simmons himself elsewhere admits, "coercion by governments is unjust...only when it extends beyond the enforcement of our natural duties to others" (Simmons 1987, 277; cf. Windeknecht 2012). Hence, I offer, the fiduciary principle can find some meaningful application, for example as a gradient of legitimate coercion, at the intersection between natural rights and duties, without requiring consent from the parties involved.

(2) If the "literalist" and the "analogical" approaches cannot be valid because of irreducible structural differences between private law and public law, which make inadmissible or impracticable the application of fiduciary legal constrains to the State, a reasoned translational approach could offer valuable indication as to a possible moral fiduciary set of duties that would constrain the holders of public offices. For example, the fiduciary fundamental obligation of loyalty in its

public declination (fairness and reasonableness), or its further derivations (such as the principle of care, the principle of deliberation, the principle of conscientiousness, and the principle of robustness; see Galoob and Leib 2018b) would offer us an intriguing "political morality that we can apply to elected rulers" (Leib and Ponet 2011, 1215). Of course, since this elaboration is not meant to impose legal obligations, but moral and political ones, political philosophers and political theorists should devise alternative ways to enforce such duties: for example through "extralegal mechanisms" such as "elections, civil society, newspapers, and watchdog groups" (Leib et al. 2013, 101).

6 Concluding Remarks

Fiduciary political theory is a promising and flourishing field of research, which brings together the efforts of law scholars, political theorists, political philosophers, and historians of political thought. In the preceding paragraphs, after a brief historical overview, I discussed a specific version of the fiduciary theory, namely, Fox-Decent's comprehensive view of the State as fiduciary of its subject. This approach is particularly interesting because it tackles many burning issues of political theory and political philosophy. I have chosen to focus on the question of political obligation, which permits a closer comparison of the fiduciary theory with its main opponent, political voluntarism and its correlate: philosophical anarchism. Like Simmons (one of the main theorists of voluntarism and philosophical anarchism), Fox-Decent ultimately bases political legitimacy, or political obligation, on a significant moral relation between the State or government and the subject or individual. While in the case of Simmons's voluntarism this moral relation is characterized by explicitly expressed consent, in the case of Fox-Decent's fiduciary theory the relation is trust-like and based on a presumption of trust. It has been shown that Fox-Decent draws on Annette Baier's notion of "unconscious" trust in order to support his argumentation. As some critics have advanced, his effort does not seem to achieve its goal. The fulfillment of the duties demanded by the fiduciary principles can give some reasons to obey the State; it does not legitimize its authority. In order to achieve full legitimacy, some kind of consent seems to be necessary. But if natural morality ultimately trumps consent, as I suggested, some action of the government can be legitimized without needing anybody's consent. Fox-Decent proves to be aware of the possibility of combining natural law and the fiduciary principle, as shown by his reflections on Thomas Hobbes (Fox-Decent 2011, 1–22; Fox-Decent 2012). This combination deserves further research.

References

Adams J (1851) The works of John Adams, vol III. Charles C, Little and James Brown, Boston
Ames JB (1908) The origins of uses and trusts. Harvard Law Review. 21:261–274. https://doi.org/10.2307/1324734
Baier A (1986) Trust and antitrust. Ethics. 96:231–260
Barker E (1962) Social contract. Clarendon Press, Oxford
Bentham J (2005) An introduction to the principles of morals and legislation. Oxford University Press, Oxford
Braithwaite J (1998) Institutionalizing trust, enculturating trust. In: Braithwaite V, Levi M (eds) Trust and governance. Russell Sage Foundation, New York, pp 343–375
Conniff J (1977) Burke, Bristol, and the concept of representation. West Polit Quart. 30:329–341
Conniff J (1993) Burke and India: the failure of the theory of trusteeship. Polit Res Q 46:291–309
de Wilde M (2011) Fides publica in Ancient Rome and its reception by Grotius and Locke. Tijdschrift voor Rechtsgeschiedenis. 79:455–487. https://doi.org/10.1163/157181911X596394
Dunn J (1984) The concept of 'Trust' in the politics of John Locke. In: Rorty R, Schneewind JB, Skinner Q (eds) Philosophy in history: essays on the historiography of philosophy. Cambridge University Press, Cambridge, pp 279–301
Finn P (1995) The forgotten "Trust": the people and the state. In: Cope M (ed) Equity: issues and trends. Federation Press, Sydney, pp 131–151
Foa Dienstag J (1996) Between history and nature: social contract theory in Locke and the founders. J Politics 58:985–1009. https://doi.org/10.2307/2960146
Fox-Decent E (2005) The fiduciary nature of state legal authority. Queen's Law J 31:259–310
Fox-Decent E (2011) Sovereignty's promise. The state as fiduciary. Oxford University Press, Oxford
Fox-Decent E (2012) Hobbes's relational theory. Between power and consent. In: Dyzenhaus D, Poole T (eds) Hobbes and the law. Cambridge University Press, Cambridge, pp 118–144. https://doi.org/10.1017/cbo9781139137034.007
Frankel T (2011) Fiduciary law. Oxford University Press, Oxford
Galoob SR, Leib EJ (2018a) Fiduciary political theory and legitimacy. In: Criddle EJ, Fox-Decent E, Gold AS, Kim SH, Miller PB (eds) Fiduciary government. Cambridge University Press, Cambridge, pp 163–182. https://doi.org/10.1017/9781108155267.008
Galoob SR, Leib EJ (2018b) The core of fiduciary political theory. In: Gordon Smith D, Gold AS (eds) Research handbook on fiduciary law. Elgar, Cheltenham-Northampton, pp 401–417
Gold AS (2013) Reflections on the state as fiduciary. Univ Toronto Law Journal. 63:655–670
Gold AS (2018) The state as a wrongful fiduciary. In: Criddle EJ, Fox-Decent E, Gold AS, Kim SH, Miller PB (eds) Fiduciary government. Cambridge University Press, Cambridge, pp 183–202. https://doi.org/10.1017/9781108155267.009
Gough J (1950) John Locke's political philosophy: eight studies. Clarendon Press, Oxford
Gvelesiani I (2016) In-depth analysis of the historical terms related to the common law "Trust". Chall Knowl Soc 6:164–169
Hampton J (1986) Hobbes and the social contract tradition. Cambridge University Press, Cambridge
Knoll M (2010) John Locke als Vordenker der Grundprinzipien des democratischen Verfassungsstaates. In: Salzborn S (ed) Der Staat des Liberalismus: die Liberale Staatstheorie von John Locke. Nomos, Baden-Baden, pp 211–244
Laslett P (1988) Introduction. In: Locke J (ed) Two treatises of government. Cambridge University Press, Cambridge, pp 3–126
Leib EJ, Ponet DL (2011) Fiduciary law's lessons for deliberative democracy. Boston Univ Law Rev 91:1207–1219
Leib EJ, Ponet DL, Serota M (2013) Translating fiduciary principles into public law. Harvard Law Rev Forum 126:91–101
Leib EJ, Ponet DL, Serota M (2014) Mapping public fiduciary relationships. In: Gold AS, Miller PB (eds) Philosophical foundations of fiduciary law. Oxford University Press, Oxford, pp 388–403
Locke J (1988) Two treatises of government. Cambridge University Press, Cambridge

Luisi N (2016) Brevi considerazioni sull'origine romanistica della struttura negoziale del trust. Studi Urbinati, A - Scienze giuridiche, politiche ed economiche 69:51–131. https://doi.org/10.14276/1825-1676.536
Maitland FW (1910) Equity: also the forms of action at common law. Cambridge University Press, London
Maloy JS (2009) Two concepts of trust. J Politics 71:492–505. https://doi.org/10.1017/S0022381609090410
Mill JS (1998) On liberty and other essays. Oxford University Press, Oxford
Miller PB (2016) Principles of public fiduciary administration. In: Kahana T, Scolnicov A (eds) Boundaries of state, boundaries of rights human rights, private actors, and positive obligations. Cambridge University Press, Cambridge, pp 251–270. https://doi.org/10.1017/cbo9781107588943
Nacol EC (2011) The risks of political authority: trust, knowledge and political agency in Locke's second treatise. Polit Stud 59:580–595. https://doi.org/10.1111/j.1467-9248.2010.00878.x
Pettit P (1997) Republicanism. A theory of freedom and government. Oxford University Press, Oxford
Pettit PH (2012) Equity and the law of trusts. Oxford University Press, Oxford
Seipp DJ (2011) Trust and fiduciary duty in the early common law. Boston Univ Law Rev 91:1011–1037
Simmons AJ (1981) Voluntarism and political associations. Virg Law Rev 67:19–37. https://doi.org/10.2307/1072830
Simmons AJ (1987) The anarchist position. A reply to Klosko and Senor. Philos Public Aff 16:269–279
Simmons AJ (1993) On the edge of anarchy. Locke, consent, and the limits of society. Princeton University Press, Princeton
Simmons AJ (2001) Justification and legitimacy. Essays on rights and obligations. Cambridge University Press, Cambridge
Windeknecht RG (2012) Law without legitimacy or justification? The flawed foundations of philosophical anarchism. Res Publica 18:173–188. https://doi.org/10.1007/s11158-011-9175-x
Zuckert MP (1994) Natural rights and the new republicanism. Princeton University Press, Princeton

Populism as the Crisis of Political Trust

Antonio Masala

Abstract Populism is an ideal litmus test to interpret the crisis of trust that today seems to affect liberal democracies and their institutions. It is by looking at this crisis that some deep motivations of populism can be understood. However, it is also by analyzing populism as a crisis of political trust that we can understand in which way the "prism of trust" is essential to understand political phenomena. This article shows how populism appears as a crisis of political trust in democracy, in representation, and in authority, and try to develop a theoretical framework about this phenomenon in terms of political theory.

Keywords Populism · Political trust · Democracy · Representation · Authority · Political order

1 Introduction

It is a widespread belief that today there is a growing crisis of political trust,[1] at least in its traditional forms (Zmerli and Van der Meer 2016). A worrisome crisis, considering that political trust is a necessary precondition of political legitimacy as well as democracy.

In many ways, populism seems to be a complete expression of this crisis. Today it appears to have become a permanent feature (and, according to many, a threat) not only in developing or weak democracies but also in countries that have a long-established democratic tradition. There are many studies on the different dimensions (political, social, economic, legal) of populism. Moreover, the same can be said of political trust, frequently discussed by recent literature in different disciplinary fields.

---

[1] I use the concept of Political Trust in the sense of the citizens' confidence in political institutions and political players. Political trust is a product of political factors and can be considered as a "one-dimensional attitude" (Hooghe 2011); it is something different from, even if strongly interconnected to, social trust (Newton 1999; Bauer and Freitag 2018), which could be indeed a prerequisite for trusting political institutions (Zmerli and Newton 2016).

A. Masala (✉)
University of Pisa, Via P. Paoli 15, 56126 Pisa, Italy
e-mail: antonio.masala@unipi.it

A. Fabris (ed.), *Trust*, Studies in Applied Philosophy, Epistemology and Rational Ethics 54, https://doi.org/10.1007/978-3-030-44018-3_13

However, quite surprisingly, the problem of how populism compares to the issue of political trust seems extremely understudied; actually, in spite of some valuable empirical analyzes (Kaase 1999; Hetherington 1999; Hooghe and Dassonneville 2016; Rooduijn et al. 2016; Algan et al. 2017), nobody has yet developed a framework of the phenomenon in terms of political theory.

This article shows how populism is also the consequence of a crisis of political trust in democracy, in representation, in authority. It is by looking at the crisis of political trust that some deep motivations of populism can be understood. However, it is also by analyzing populism as a crisis of political trust that we can understand in which way the "prism of trust" is essential for understanding political phenomena; though widely investigated, such topic still has precious heuristic value.

2 Populism, Trust, Democracy

Populism is an elusive and hardly definable political phenomenon; in particular, it is not easy to identify theoretical constants that go beyond the historical cases. A minimalist definition of populism could be: a political proposal that rejects any mediation in order to restore the sovereignty of the people (taken away by the elites) and do its will. In this sense, the populist phenomenon appears as a crisis of trust in democracy, in two ways. On the one hand, it is a crisis of confidence in liberal democratic institutions, which should be replaced by a more direct and unmediated democracy. On the other hand, it is a crisis of confidence in the ruling classes (that the populists always call establishment or elite), regardless of their political party, who never really wanted to fulfill the will of the people. The two dimensions are interconnected.

One of the reasons for populist consent is the tendency of many democracies to expand the government's responsibilities and fulfill the citizens' expectations. This process often led to rule through a sort of commutative contract, whereby economic benefits are provided in return for consent—a critical problem, which may bring corruption into the democratic institutions (della Porta and Vannucci 2012) and has important consequences in order to understand the relationship between populism and liberal-democracy. In the long term, this trend would result in a loss of confidence, not only in the ruling class but also in the institutions (which are no longer perceived as impartial). Since promising favors is easier than building institutional trust, the consequence would be the deterioration of the political offer. In this sense, populism seems, on the one hand, the result of bad policies and the consequent loss of trust in politics; and, on the other hand, the sublimation of this process, because populist leaders lead us to believe that, once the old elites (who are pursuing personal interests) are replaced, benefits can be shared by everyone.

Then, populism is a call for trust in a new politics, based on the de-legitimization not only of other politicians, but also of democratic institutions. The lower the level of trust in institutions, the greater the tendency to choose the populist political *offer* that promises more immediate and personal benefits, as opposed to good rules for the community. Similarly, in this sense, populism can also be seen as a potential *supply* of trust from populist leaders to an indistinct people, substituting other forms of institutional trust.

In order to analyze such dimensions, it is essential to define the populist ideology, or, better, to ask ourselves if a populist ideology actually exists and, above all, how it is configured. Defining populism as an ideology is not easy. Those who tried did so by talking about a "thin-centered ideology" (Canovan 2004, 32; Mudde 2004; Mudde and Kaltwasser 2017) or a "weak ideology" (Mény and Surel 2000). These are consistent attempts which, however, do not seem to solve the problem of finding a permanent, consistent content, which all ideologies have, for the different forms of populism.

If we understand ideology as a combination of beliefs built on ideas and values, and as their manifestations in political practice, defining populism as an ideology can be quite difficult. There are no founding fathers or textbooks in populist ideology (Tarchi 2015: 40), and there does not seem to be a steady core of basic values similar to the one that exists in other political ideologies/theories. If liberalism has individual freedom as its guiding principle, democracy has political equality, socialism has economic equality, no such values can be identified in a hypothetical populist ideology; no actual "content" can be identified as the inspiration of populism (Taguieff 2008).

Fulfilling the will of the people may perhaps be regarded as the underlying political value of populism. However, it is a different type of value from those that give substance to political ideologies. The people may want very different, even opposite, things in terms of values. In this sense, the different forms of populism endorse opposing views of how political life should be organized.

Perhaps, the essence of populism lies in the promise that one can actually achieve something that is not really possible, in considering the people as something real and identifiable, and democracy not "just" as a right to choose representatives who can make complex choices (Müller 2016: 76); such view of populism seems incompatible with a liberal interpretation of the voting process as mainly an instrument to remove unpopular politicians (Riker 1982, 241–246). Populism hypostatizes the people as a unitary (if not organic) entity (Urbinati 2019; Mény 2019), but people do not really exist in such terms, and then they must be artificially constructed. In all populist practices, there is in fact an attempt at "constructing" people, and excluding those (the "corrupted") who do not support populists is a distinctive feature of populism (Akkerman et al. 2014). In most cases, people are constructed by instilling fears and by finding an enemy of the "real people", who is often, but not always, considered an ethnic entity. The enemy can be the establishment that holds power and has betrayed the people, using it for its own benefit, not for the common good. However, the enemy can also be the immigrants, the minorities, or the "different ones". Fear, if not hate, of those who do not belong to the people is, therefore, a typical feature of populism, and it is also what makes it appear as a politically worrisome and undemocratic phenomenon, one that even be willingly used o undermine democratic systems.

Many analyses of populism deal with the concept of the people and the tensions between populism and liberal democracy. The belief that the people are not and cannot be a "physical" entity, with a uniform will, is at the heart of modern liberal democracies, which rely on the concept of popular sovereignty as the product of a representative mechanism. Populism represents the denial of that concept of democratic sovereignty, which must necessarily pass through the mechanism of representation and the limits set by the constitution (Sartori 1962; Urbinati 2015).

Populists present themselves as intending to accomplish a "true" democracy, which identifies with the will of the people; they do not criticize democracy itself, but the way in which it has been expressed and betrayed in actual fact by those who should have fulfilled the will of the people. The idea of a betrayed democracy and the controversy over the unfulfilled promises of democracy became a weapon of propaganda, regardless of the values of the populist premise. The tension between the ideology and the practice of democracy becomes the "cry of democracy betrayed", and a tool to mobilize the discontented to restore politics to the people (Canovan 2004: 26).

The relationship with democracy is central to any understanding of populism. We have already seen how populism has no classic textbooks or founding fathers, but who are the classics of democracy? Among them are the ancient Greeks and Rousseau, who left us some founding elements of democratic theory, but, undoubtedly, their view of democracy has very little to do with today's concept, which in some respects resembles the populist view, instead.

If we look at the great contemporary works on democracy, we can see how many of them warn us against the dangers of a democratic theory. Joseph Schumpeter, to name just one, maintains that the basis of the classical democratic theory (which is not that of the Greeks but that of Jeremy Bentham) is the belief that there is a common good capable of "offering precise answers to every question", and that it can be achieved by consulting the popular will. However, if democracy wants to be the actualization "of the popular soul", its results can be very different from one another, and even very far from those ideals that we commonly attribute to democracy today (Schumpeter 1942).

We could then say that, when populism hypostatizes the people and proposes to be the realization of the soul of the people, it turns into an ideological view of democracy (in the worst sense). In fact, populism is characterized by a "propagandistic" and instrumental use of democracy, which is understood only as the power of the people and as popular sovereignty. It absolutizes such elements of democracy and rules out the others, who are also an integral part of it: first of all, the concepts of pluralism and the respect for minorities. It then turns into demagogy—which, according to the ancient Greeks usually goes along with a degenerated form of democracy—and can in many ways be analyzed as a method for conquering power, regardless of the project and values that populist leaders want to achieve (Taguieff 2008).

This form of populism is driven by a logic that is not the one of liberal democracy, where even "deep political disagreements" are considered as normal and (usually) fruitful (Knoll et al. 2018). If liberal democracy advocates the protection of minorities that can become political majorities tomorrow, populism is just the opposite, because minorities are not considered part of the "people". This type of populism, which is not the only possible one and which we could perhaps define as "ideological populism", tends to occur mainly in young or non-mature democracies and often brings about authoritarian regimes. However, there is no denying that ideological populism is a challenge that, in recent years, seems to have regained strength, even in countries with a long-standing democracy.

To fully understand the challenge ideological populism poses to liberal democracy, let's look at the way universal suffrage is differently conceived. In liberal democracy, universal suffrage is not only the point of arrival of a historical path that leads to the gradual extension of the right to vote. Universal suffrage is also a point of arrival in terms of political theory; in fact, to ensure the proper functioning of the democratic system, some civil rights are needed, and those rights are the very essence of democracy. It is not just about the right to vote and the counting of votes, it is about the conditions on which we get to vote: freedom of expression and information, ease of access for new political players, protection of minority rights, and more.

Conversely, for ideological populism, universal suffrage is only the starting point of a path that aims to achieve popular sovereignty, which continues even after voting; it is the mirage of having people in control of the real and perpetual, which is essentially impossible in modern democracies (Achen and Bartels 2016). Populism does not look at the conditions of freedom that must exist before voting (and be protected after voting) and which guarantee that society can maintain its plurality. In populism, the view of politics as mediation and discussion disappears, a view that only makes sense if the expression of popular sovereignty is only allowed at election time, and cannot be called up afterwards, unless in exceptional circumstances.

The need to revive the popular will beyond the time of the vote expresses a deficiency of political trust. The populist voters want, first of all, to send out a highly critical message about the poor functioning of the political system and want to express their lack of trust in it (Hooghe and Dassonneville 2016).

3 Populism, Trust, Representation

Populism, in a nutshell, can be construed as a symptom of the crisis of trust in politics (as well as in democracy); it relies on the assumption that elites and institutions do not know how to meet the needs and fulfill the will of the people. The populist premise sees the solution of this "crisis of trust" in the establishment of a new ruling class (and often in a charismatic leader) that breaks with the abstruse mechanisms of representative democracy in order to achieve what people really want.

Although the need for representation is never ultimately denied, in populism the will to reform it is always present; the populist aim is a system where the representatives are *obliged* to take into account the wishes of the represented. The real problem of politics is the distance between the ruling class and the people, its "diversity" from the people. The establishment does not solve people's problems because it pursues personal interests and is probably not even interested in knowing what people really want. Populism is in many ways a massive operation of simplification: the problem of politics does not lie in complex choices and in the tension between them, but in the bad faith of politicians who do not want to pursue the interest of the people.

According to populists, the people (as an organic entity) know what they want and accomplishing that is only a matter of willingness. If politicians are honest and listen to people, accomplishing what they want is not a problem; the difficulty is not in the accomplishment, it is in not listening to people's voice.

This is why in populism the dynamics of trust often moves along the simplistic line of identification, of similarity. I trust those who are similar to me; I choose those who look like me as my representatives, because they know what I need and I can trust them, and only them. Politicians, and decision-makers in general, must be "men of the people" because the difficulty is only in knowing what the people want and being honest in order to achieve it. Quite paradoxically, something similar to this mechanism of identification seems at work when a populist leader is not a "man of the people" but is, for example, a billionaire or a celebrity. In this case, again, the new political leader is something different and far from the "old" politicians, someone who can offer a new story and a new view of politics, someone who is not as corrupted as professional politicians are, and, as such, can be trusted to offer a political project in the interest of the people.

In populism, the idea of representation as a complex phenomenon breaks down. The concept—whose classical definition goes back to Edmund Burke—that sees representation primarily as a "question of reason", and the problem of representation as the problem of choosing those who better than others are able to solve the complex problems of living together, basically disappear. Burke's idea was that, when choosing representatives, one must look for someone who is not equal to himself but better, more competent, more capable of solving political problems that are complex problems. At the root, there is certainly a view of the people as something plural, which has different and sometimes opposing interests, but also the conviction that political choices are complex choices, choices that can have unpredictable outcomes, and knowing what you want to see achieved with simple choices is not enough, therefore.

This element is particularly relevant: in populism—or at least in some forms of populism—the problem of competence, of making laws as a matter of reason, disappears. Once a problem has disappeared, only the matter of willingness remains: we need to know what the people really want, then we can easily fulfill their will. A political leader has to be a man who knows what the people want because he belongs to the people. A leader who speaks the language of identification and builds political trust upon it. The loss of confidence in the ruling classes—as the purpose of looking for representatives (and rulers) who are better than us—is perhaps the most significant feature of populism and is the litmus test of the current political crisis. A crisis that, according to some scholars, sees representative democracy suffering precisely because of a lack of political trust. A lack that has serious consequences on the quality of democratic players and institutions (Dalton 2004, Chap. 8; Thomassen 2015).

The mechanisms of political trust (and therefore of consensus building) in populism are therefore different from those of trust in traditional representative systems. We trust the populist leader also because (or precisely because) we no longer trust the institutions, the collegiate bodies of representation, and because we no longer trust those ideologies that proposed clear views of the world. The populist leader has a qualitatively different trust—built on similarity or admiration when the leader is successful in a field other than politics—rather than political qualities. It is a personal trust, built not so much, or not only, on a proposed solution, on a political program, but on non-political elements.

The populist leaders' communication tells a lot in this respect. It involves oversimplified, easily understandable political messages, and also portraits the politician as a person, often with habits and tastes similar to those of the people. Populist politicians show where they go on holiday or to relax, or what they eat and like, so they can be felt as more authentic than the "old" politicians. In this sense, the growing importance of social media becomes a key tool for populists to build political consensus (Moffitt 2016, Chap. 5; Enli and Rosenberg 2018). However, perhaps the most interesting aspect is how this construction of political trust (also) involves contents that have nothing to do with the political program.

Not only are some important elements connected with the change in the communication system that is strongly encouraged by social media. There's no denying that the building of trust and political consensus has been going on for some time, starting with an extraneousness to politics and (at least in part) to the contents of politics. The deficit of trust in the political system often rewards proposals that are not labeled according to the traditional categories, and successful political leaders are those who have not been engaged in politics before or who are professional politicians who speak the language of anti-politics or at least the language that criticizes "old" politics. The apparent paradox is that presenting oneself as distant from an allegedly abstruse and overcomplicated politics and proposing a political program that aims to fulfill the will of the people against politics becomes the element on which a new political trust should be built.

This process goes hand in hand with that of personalization in politics, a process that seems to have been remarkably accelerated by populism. Identifying with a person is easier than identifying with a political program, with institutions, and with procedures. An energetic leader seems a more suitable solution to overcome the complexity of the political mechanisms, often blamed for the failure to achieve the popular will. In a sense, populism is a personalization of politics: political trust is based not upon programmatic proposals but upon the personal appearance and communication skills of the leader, who wants to be felt as a quasi-familiar person. Political trust in this kind of leader is very different from political trust in the institutions, and it seems to be a failure of the complex mechanisms of representative democracy.

4 Populism, Trust, Authority

However, populism also seems to be a litmus test to analyze a much more generalized crisis of trust, which affects not only the representative institutions but also almost every institution of any democratic system, whether elective or not, to the point of being a crisis of trust in any form of authority.

Contemporary liberal democracies, which, not by chance, are also known as polyarchies (Dahl 1971), work by balancing different centers of power, some of which are independent of popular investiture, or have received such investiture in a very mediated and indirect way. Examples of this are the judiciary, bureaucracy and, increasingly, some supranational institutions. The plurality and independence of these powers give strength and substance to democracy, understood as liberal democracy. In the populist's eyes, however, such powers prevent the accomplishment of the will of the people, as they do not meet popular sovereignty, and belong therefore to what populism identifies as the enemy, the establishment. If, according to the populist view, the people are one and have one will, imposing limitations on them makes no sense, especially if the institutions that impose such restrictions are not a direct expression of the popular will. These are the causes of a crisis of trust in the institutions that act in an impersonal manner and follow established rules rather than the "popular will". It is a crisis of confidence in the authorities that do not have a popular investiture, which is opposed to by a proposal to trust those politicians who can be identified as belonging to the people and who want to fulfill the people's demands.

However, populism becomes the embodiment of the crisis of trust in authority, even in a deeper sense. To understand this, we can look at how some political and social institutions have emerged and established themselves. The tradition of classical liberalism (Menger 1883) has shown that some of the main institutions that serve human well-being are not the product of a deliberate choice, of a political will aimed at that result. This tradition has shown how human actions, first and foremost collective choices, always give rise to a series of unexpected and sometimes undesired consequences. This discovery is, first of all, significant to expose the limits of politics. Not all the best solutions for civil coexistence are the result of a will, of a deliberate choice; some of them stem from a process that is largely spontaneous, or at least unintentional, which leads to recognize some solutions as better than others and to adopt them, or conform to them. In this sense, trust in institutions and authorities is aroused not only by electoral processes and consensus by majorities, but by the positive results they deliver; it is a trust based on authority, not on voting. In many ways, the process is similar to what happens in the world of research (where scientific theories succeed because they are correct, not because they are "popular") and in the market (where the products that succeed are those that best meet the consumers' needs). This process has a long history in the field of law (which, in the Roman world and then in the Anglo-Saxon tradition, was not only the one created by parliaments but the one made by magistrates) and in many other cases as well (Leoni 1961).

In these circumstances, the trusting mechanism is fragile and complex. It involves the acceptance of rules and results that were not chosen, and which can sometimes disappoint. Consider, for example, the functioning of the free market. It can deliver results that are considered to be subjectively unfair, but thinking that the market should only be ruled by political choices based on electoral consensus would negate its ability to efficiently meet the consumers' needs. It is, therefore, trust in mechanisms, institutions or authorities that do not depend on human will that are conducive to the accomplishment of what liberal political philosophers call a good political order.

Populism calls into question any possibility of trusting all this, fostering the idea that all forms of authority must be ascribable to the people, and that nothing must be alien to the will of the people. Once the will of the people has been expressed, then we can give society the form we want, and, in this sense, populism seems to become a variant of what Friedrich A. von Hayek called constructivism (Hayek 1952). Nothing is precluded to what the (populist) politician can accomplish; as long as he knows how to understand what the people want, realizing it with a political action will not be a problem.

What is commonly known as "sovereignism" is an expression of populist mentality. The people, the national community, must decide. The international institutions, the markets, the processes of globalization become the enemies that want to take away from the people the possibility of deciding for themselves. The construction of political trust then moves according to the dynamics of recognition and closure; "I give confidence to those I know, to those who are like me, to those who I can control with my vote". A dynamic of trust that sounds archaic and that has the erosion of the "culture of trust" as its counterpart (Sztompka 1998): the crisis of trust in the complex mechanisms and institutions of the open society. A crisis that is based on the fear and rejection of those phenomena and institutions that are not the product of a deliberate choice and that do not seem completely controllable by human will.

5 Conclusion

This article has demonstrated how populism is an ideal litmus test to interpret the crisis of trust that liberal democracies and their institutions seems to be permeated with, nowadays. Political representation and independent institutions reflect the need to achieve democracy through intricate paths, which take into account the popular will expressed at election time while still submitting the sphere of political action to the rules and procedures that govern it. Trust in these rules and institutions, and in their independence, is trust in democracy; populism negates representative democracy in many different ways and, by bringing every decision back to the will of the people (a hypostatized, non-real people), is an expression of a generalized crisis of political trust.

The reasons for this crisis are certainly many and varied, some of them related to contingent political failures, others to structural difficulties. Among the latter is the fact that politics has been overloaded with tasks and expectations (Orsina 2018). In liberal democracies, for a long time politicians have suggested that all individual expectations, or at least many of them, might turn into rights, which politics had to fulfill, and that the task of politics is to satisfy an ever-increasing number of human desires. The thought that such a task was feasible (and was not) generated a widespread crisis of trust in the institutions of democracy and in politics itself.

However, this generalized crisis of political trust does not lead to a call for a restriction on political activity. Instead, it has often led to the spread of populism, which does nothing but promise to fulfill (through politics) the growing desires of the people, desires which, the populists claim, have not been met because someone stands in the way of the popular will. Therefore, the political proposal of populism always seems to go along with discrediting what is referred to as the "opaque" institutions of democracy, and with all the phenomena that escape the direct control of the will of the people. The consequence is that political trust is increasingly moving along the simple, or rather simplistic, dynamics of the similarity between the leader and the people, and less and less on the much more intricate one of trust in institutions and in complex political proposals.

We live in a society of communication technology and disintermediation. In this society, the possibility of expressing personal opinions and the conviction (or perhaps the illusion) of making an unfiltered contribution to the decision-making processes are expanding; therefore, the need for trust is increasing, but in a very different way from the past (Fabris 2020). The new populisms, by leveraging on social media, convey a feeling of closeness between the leader and the people, as well as the illusion of rebuilding the lost social cohesion. It is, therefore, the ambiguous and dangerous attempts to deny the "old" mechanisms of trust and to replace them with new (and technological) trust-building dynamics that seem to be more superficial and instinctive. In this respect, populism looks like a decline in the quality of political trust and in politics itself.

References

Achen CH, Bartels LM (2016) Democracy for realist. Why elections do not produce responsive governments. Princeton University Press, Princeton and Oxford

Akkerman A, Mudde C, Zaslove A (2014) How populist are the people? Measuring populist attitudes in voters. Comp Polit Stud 47(9):1324–1353

Algan Y, Guriev S, Papaioannou E, Passari E (2017) The European trust crisis and the rise of populism. Brookings papers on economic activity, fall, pp 309–382

Bauer PC, Freitag M (2018) "Measuring Trust". In: Uslaner EM (ed) The Oxford handbook of social and political trust. Oxford University Press, pp 15–36.

Canovan M (1999) Trust the people! Populism and the two faces of democracy. Polit Stud 47(1):2–16

Canovan M (2004) Taking politics to the people: populism as the ideology of democracy, In: Mény Y, Surel Y (eds) Democracies and the populist challenge. Palgrave, London

Dahl RA (1971) Polyarchy: participation and opposition. Yale University Press, New Haven

Dalton RJ (2004) Democratic challenges, democratic choices: the erosion of political support in advanced industrial democracies. Oxford University Press, Oxford

della Porta D, Vannucci A (2012) The hidden order of corruption. An institutional approach. Ashgate, Aldershot

Enli G, Rosenberg LT (2018) Trust in the age of social media: populist politicians seem more authentic. Social Media + Society, pp 1–11

Fabris A (2020) Populismi, comunicazione, fiducia. In: Masala A, Viviani L (eds) L'età dei populismi. Carocci, Roma

von Hayek FA (1952) The counter-revolution of science: studies on the abuse of reason. Free Press, Glencoe Ill
Hetherington M (1999) The effect of political trust on the presidential vote, 1968–96. Am Polit Sci Rev 93(2):311–326
Hooghe M (2011) Why there is basically only one form of political trust. Br J Polit Int Relat 13(2):269–275
Hooghe M (2018) Trust and elections. In: Uslaner EM (ed) The Oxford handbook of social and political trust. Oxford University Press, Oxford, pp 617–631
Hooghe M, Dassonneville R (2016) A spiral of distrust: A panel study on the relation between political distrust and protest voting in Belgium. Gov Oppos 1:1–27
Kaase M (1999) Interpersonal trust, political trust and non-institutionalised political participation in Western Europe. West European Politics 22(3):1–23
Kenneth N, Stolle D, Zmerli S (2018) Social and political trust. In: Uslaner EM (ed) The Oxford handbook of social and political trust. Oxford University Press, pp 37–56
Knoll M, Simsek N, Snyder S (eds) (2018) New Perspectives on distributive justice: deep disagreements, pluralism, and the problem of consensus. De Gruyter, Berlin
Leoni B (1961) Freedom and the law. Van Nostrand Company Inc, Princeton
Menger C (1883) Untersuchungen über die Methode der Socialwissenschaften, und der Politischen Oekonomie insbesondere. Duncker & Humblot, Leipzig
Mény Y (2019) Popolo ma non troppo. Il Mulino, Bologna
Mény Y, Surel Y (2000) Par le peuple, pour le peuple: le populisme et les démocraties. Fayard, Paris
Moffitt B (2016) The global rise of populism. Performance, political style and representation. Stanford University Press, Stanford California
Mudde C (2004) The populist zeitgeist. Gov Oppos 39:542–563
Mudde C, Kaltwasser R (2017) Populism. A very short introduction. Oxford University Press, Oxford
Müller J-W (2016) What is populism. University of Pennsylvania Press, Philadelphia
Newton K (1999) Social and political trust in established democracies. In: Norris P (ed) Critical citizens. Oxford University Press, Oxford, pp 169–187.
Orsina G (2018) La democrazia del narcisismo. Marsilio, Venezia
Riker WH (1982) Liberalism against populism: a confrontation between the theory of democracy and the theory of social choice. Waveland, Long Grove (Illinois)
Rooduijn M, Van der Brug W, De Lange S (2016) Expressing or fuelling discontent? The relationship between populist voting and political discontent. Electoral Stud 43:32–40
Sartori G (1962) Democratic theory. Frederick A. Praeger, New York
Schumpeter JA (1942) Capitalism, socialism and democracy. George Allen e Unwin, London
Sztompka P (1998) Trust, distrust and two paradoxes of democracy, . Eur J Soc Theory 1–1:19–33
Taguieff P-A (2008) L'illusion populiste. Essai sur les démagogies de l'âge democratique. Flammarion, Paris
Tarchi M (2015) L'Italia populista. Il Mulino, Bologna
Thomassen JJA (2015) What's gone wrong with democracy, or with theories explaining why it has? In: Poguntke T, Rossteutscher S, Schmitt-Beck R, Zmerli S (eds) Citizenship and democracy in an era of crisis. Routledge, London
Urbinati N (2015) Democracy disfigured. Opinion, truth and the people. Harvard University Press
Urbinati N (2019) Me the people. How populism transforms democracy. Harvard University Press
Zmerli S, Newton K (2016) Objects of political and social trust: scales and hierarchies. In: Zmerli S, van der Meer TWG (eds) Handbook on political trust. Edward Elgar Publishing, Chelton, U.K
Zmerli S, Van der Meer TWG (eds) (2016) Handbook on political trust. Edward Elgar Publishing, Chelton, U.K.

Author Index

A. Fabris (ed.), *Trust*, Studies in Applied Philosophy, Epistemology and Rational Ethics 54, https://doi.org/10.1007/978-3-030-44018-3

The manufacturer's authorised representative in the EU is Springer Nature Customer Service Centre GmbH, Europaplatz 3, 69115 Heidelberg, Germany. If you have any concerns regarding our products, please contact ProductSafety@springernature.com

Printed and bound by CPI Group (UK) Ltd, Croydon, CR0 4YY
30/09/2026
02225645-0015